Dynamics in Economic Geography

Dynamics in Economic Geography

Changing views on industrial locations and regional development

Oedzge Atzema
Ton van Rietbergen
Jan Lambooy
Sjef van Hoof

First English language edition

uitgeverij coutinho | c

bussum 2016

First edition 1997, Revised second edition 2002, Revised Third edition 2012
First English language edition 2014 based on the Revised Third Dutch language edition 2012, second printing 2016

Uitgeverij Coutinho
Postbus 333
1400 AH Bussum
info@coutinho.nl
www.coutinho.nl

Cover design: Ronald Boiten, Amersfoort
Cover photographs: © Universiteit Utrecht, faculteit Geowetenschappen, C&M
Translation: Myra Wilkinson-van Hoek and Sheila Gogol

ISBN 978 90 469 6270 1
NUR 903

Foreword to revised third edition

The global economy is in a near-permanent state of flux. Banks merge and split up again, former developing countries account for an ever-growing slice of world trade and innovations force one industry into a defensive position and allow another to flourish. The integration of the European economy is stagnating and EU expansion and indeed the EU itself are increasingly called into question.

As an empirical applied science, economic geography can help explain this growing economic turbulence. Rooted in 19th-century theory, it has evolved and adapted over time. In recent decades, economic geographers have studied globalization, regional institutional variation and the spatial evolution of industries, and people are listening to what they have to say.

The aim of this book is to give students and other interested readers an up-to-date introduction to economic geography. It purports to be no more than an introduction, although the bibliography provides a good overview of the current research results. Compared with the second edition published in 2002, how the chapters are organized and the topics that are addressed have been drastically updated.

The book consists of two parts. An introductory chapter outlining the relation between economic geography and economic science is followed by four chapters addressing locational issues from a variety of perspectives revolving around why businesses choose a particular location. The next four chapters each focus on specific regions and nations and how economic differences between regions come about and can be compared.

All the chapters have the same structure, whether from a business or a regional perspective. Chapter 2 discusses the classical and neoclassical perspective, Chapter 3 the behavioural approach and Chapter 4 the institutional aspects, i.e. how industries/regions are organized. Lastly, we examine the evolutionary approach, which has recently had a sharp rise in popularity in economic geographic literature.

We would like to express our gratitude to Margot Stoete from the Faculty of Earth Sciences at the University of Utrecht for producing the figures and to Suzanne Loohuis and Jan Scheffers at Coutinho Publishers, without whose insightful comments and unwavering commitment this book would not have been feasible in its current form. We hope the book will serve as a suitable springboard into the field of economic geography for students, fellow academics and other interested readers. If you have any queries or comments, please do not hesitate to contact us.

Oedzge Atzema, Ton van Rietbergen, Sjef van Hoof and Jan Lambooy
Utrecht, Spring 2012

Contents

1

Application possibilities for economic geography

This book on economic geography addresses the main theoretical strands at the intersection of economics and geography. Economic geography has been around for centuries and has proven to be a useful discipline, whether it focuses on trade in colonial times or the current Internet era. At its most succinct, the question an economic geographer asks is 'What happens where and why is it happening there?' He devotes particular attention to spatial differences in costs and benefits between nations, regions and cities. The global imbalance in wealth distribution is a typical economic geography topic. Using data like those from the World Bank, economic geographers explain such imbalances by examining the distribution of raw materials and factors of production such as labour, capital and land. Factors affecting the accessibility of regions, such as harbours and roads, are also important, as are the scale and composition of industrial activity. The productivity of a region, and thus its ability to generate income, varies per industrial sector. In addition to economic factors, cultural, social, political and natural ones influence global wealth distribution.

Though economic geography rests on a theoretical foundation, it is an applied science. This means economic geography knowledge has to be 'useable' by government and industry. Two questions are at the heart of economic geography:

- Why does a company establish itself in region A and not in region B (i.e. factors that influence the choice of location for industry, utilities and households)?
- Why are some countries, regions or cities economically more successful than others (i.e. factors that influence the development of countries, regions or cities)?

These two questions underlie the structure of this book. In Chapters 2 to 9, they are addressed from a variety of perspectives: neoclassical, behavioural, institutional and evolutionary. This chapter gives a historical overview of the relations between economics and geography. The tenth and final chapter focuses on the current state of economic geography.

1.1 Rediscovery of economic geography

In 2009, the *World Development Report* published annually by the World Bank was entitled *Reshaping Economic Geography.* The report devoted particular attention to the role played by the physical environment. It is clear from this and from the title that geographic factors are of great importance to the World Bank in explaining economic differences between nations and regions. Another sign of increasing interest in the influence of geography on economic issues was the 2008 Nobel Prize for Economics awarded to Paul Krugman for his analysis of trading patterns and locations of economic activity.

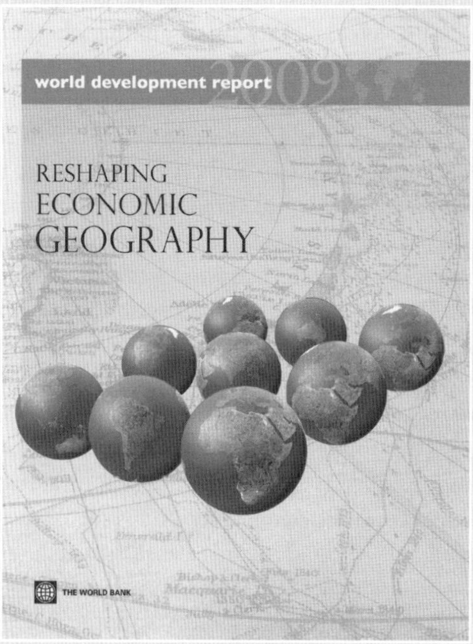

Figure 1.1
The 2009 World Development Report

Having been brought into disrepute by the Nazis, who used geographic theories to underpin their plans for the expansion of the Third Reich, the academic discipline of geography had a poor reputation after World War II. Geography was associated with *Blut und Boden* (Blood and Soil, the alleged connection between geography and race) and the geography faculties at Harvard in Boston and Columbia in New York were closed during this period, as were those at other renowned universities in the United States. For economic geography as a discipline, this was a significant blow.

Slowly but surely, economic geography recovered and certainly from 1990 onwards, it has been a flourishing discipline, though the debate between economists and geographers goes on. Economists accuse geographers of insufficient general theory, and geographers claim economists do not devote enough attention to empirical evidence as a source of knowledge. Often the debate centres on the extent to which the physical environment determines the economic achievements of businesses and regions. Over time, various answers to this question have been put forward. Initially, the focus was on the influence of the physical environment on regions' economic achievements. In his seminal work *Civilization and Climate* (1915), Ellsworth Huntington (1876-1947) classified civilizations hierarchically. The most stimulating climate, he argued, was in New Haven, Connecticut, which happened to be where he taught at Yale University. Then came other regions, with tropical countries inhabited by dark-skinned people bringing up the rear. Huntington considered climatic differences relevant to environmental factors influencing businesses, and thus overall economic achievements. Geographer Ellen Churchill Semple went further, witness the first sentence of her 1911 book *Influences of Geographic Environment*: 'Man is a product of the earth's surface.'

Huntington and especially Semple were clear representatives of physical determinism in geography based on the premise that nature limits variations in human behaviour. Economists thought this was too dogmatic and did not leave enough room for human agency, leading to a rejection of geography's ability to explain world poverty. The solution to poverty would have to come from humans and technology, for example by stimulating the economy, as was argued by Keynes and applied by Roosevelt in his New Deal to cope with the Depression.

Initially, the post-war emergence of the welfare state did not do the discipline of economic geography much good either. Society was heavily influenced at the time by the rise of new technology. There was a strong feeling among politicians, entrepreneurs and consumers that nature and the limitations it imposed could be controlled. Society was something that could be shaped into whatever people wanted it to be. This led to large-scale interventions in the physical environment, such as massive deforestation, the widespread use of pesticides and the diverting of great rivers – not only in totalitarian countries such as the Soviet Union and China but also in democratic ones. The ecological results of these interventions were often disastrous.

The late 1960s witnessed renewed interest in economic geography. This was not so much because the natural environment was considered important but because the realization was dawning that imbalances between regions may well be permanent – not only globally but also within countries. Some regions had a much higher unemployment rate than others, and the imbalances persevered. In the agricultural and industrial regions of the Netherlands for example, unemployment was particularly high at the time, prompting many people to leave for Canada, Australia and New Zealand. The tide could be turned by strengthening the local economy. In those days, regional planning policies aimed to 'bring work to the people' and the government offered grants to attract new businesses to the region or expand existing ones. In addition, access to these regions was improved by constructing roads and railways. In other words, the regional geography was adjusted to stimulate the economy and boost prosperity. So there was an economic geography application in government policy.

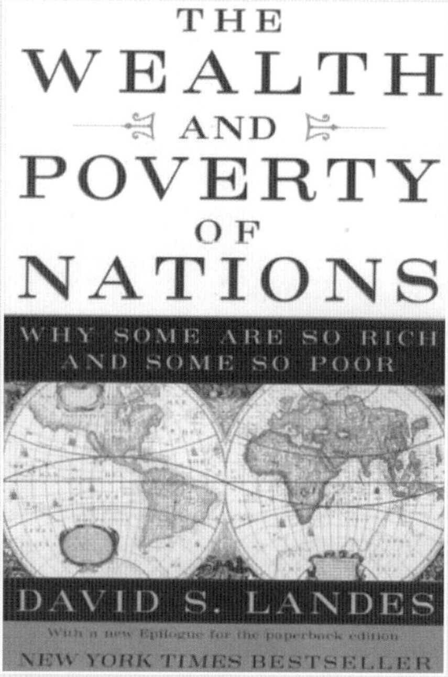

Figure 1.2
David S. Landes' bestseller *The Wealth and Poverty of Nations*

Published in 1998, *The Wealth and Poverty of Nations* by historian David Landes marked the launching of the rehabilitation of economic geography in the social sciences in the United States. Landes felt that with its emphasis on the unequal distribution of natural resources and climatic limitations, geography had wrongly become a forgotten discipline. He gave the example of a community's struggle against the sea as a possible basis for economic welfare. However, due to widespread illness, tropical climates tended to lower productivity and thus constituted a hindrance to development. Although they were grateful to Landes for putting geographic factors back on the agenda, many geographers also felt he went too far, as his ideas reminded them too much of the earlier determinism. This said, Landes undoubtedly did much to generate interest in economic geography.

While Landes emphasized the importance of the geography variable to explain differences in welfare levels, economic geographers and historians generally concentrated on how societies were organized in terms of their legal system, education and entrepreneurial climate. An apt illustration was an anecdote by Swiss economic historian Paul Bairoch, who once wrote, 'They are small countries, surrounded by often hostile neighbouring countries, without access to the sea and with no natural resources. Much of the land is mountainous and infertile. Roads are hard to construct due to the mountainous terrain. (...) The population is small and consists of several groups, without cultural or linguistic connections. People are so poor that they leave the country or sign up as mercenaries in foreign armies. The prospects of such a country would appear to be very poor indeed – the developing nation *par excellence*.' According to Bairoch, this was a fairly accurate description of Switzerland around 1800. Today Switzerland is one of the wealthiest nations in the world. Its open-door policy, tax system, education, frugality, calm political situation and neutrality are some of the factors thought to have stimulated its economic development.

Breaking the vicious circle of poverty is possible, historians argue, but it is a complex process. Jan Luiten van Zanden, Professor of Global Economic History at the University of Utrecht in the Netherlands, reduced this complexity to two factors: the division of power and access to natural resources. 'Once a growth nucleus has developed (e.g. the late medieval Italian city states, the cities in 17th-century Holland, or post-World War II Germany and Japan), such a nucleus often pulls surrounding regions along with it. Soon, "the neighbours" do more than simply bob along on the waves of trade and productivity. Often the surrounding areas copy the institutions of an economic centre. As well as reflecting economic fluctuations, this is a fundamental, structural development' (*NRC Handelsblad*, 21 June 2011).

In addition to historians, we owe the rediscovery of economic geography to economists. In *The Competitive Advantages of Nations* (1990), bestselling American business strategy expert Michael Porter described the basic geographic factors necessary for a country to be competitive. In the context of the Netherlands, these geographic factors would be the mouths of the main rivers the Dutch 'delta economy' is based on. Boasting Rotterdam harbour and Amsterdam Airport Schiphol ('the Gateway to Europe'), the Netherlands has a competitive advantage that can be directly linked to its geography. Porter gave examples of the economic success stories of regions such as the Third Italy in northern Italy and Silicon Valley in the United States.

Porter argued that nations are well advised to focus on their regional economic specializations (clusters) as the driving force behind their economic development. In the Netherlands, this line of thinking underpinned *Pieken in de Delta*, a long-running development programme focussed on regional strengths that has led for example to the establishment of a 'Food Valley' in and around the town of Wageningen. What is striking here is that we are once again dealing with an approach to economic geography aimed at policy development.

In this context it is relevant to mention Richard Florida, author of *The Rise of the Creative Class* (2002). Since manufacturing was increasingly relocated from the Western world to low-wage countries, his argument was that creativity was becoming the main driving force behind economic growth (we will expand on this in Chapter 7).

The rise of the Internet prompted O'Brien (1992) to predict the demise of geography. After all, if distance is no longer an issue, how can geography possibly have anything left to offer? Books like the provocatively titled *The World is Flat* by Thomas Friedmann (2005) seemed to confirm the picture painted by O'Brien. However, in his inaugural address, Robert Kloosterman, Professor of Economic Geography and Planning at the University of Amsterdam, cited *The Economist* to illustrate why this picture is flawed. 'Distance is dying, but geography, it seems, is alive and kicking' (Kloosterman 2001). Six years earlier, *The Economist* had proclaimed that distance was no longer relevant as a factor limiting economic activity.

Whether the world becoming a smaller place will cause the demise of geography is a matter of intense debate. In an article entitled 'Prisoner of Geography,' in 2002 the authoritative American journal *Foreign Policy* recognized that the significance of geographic factors may be underestimated in efforts to explain national differences in wealth, concluding that 'In the academic arena economic geography is no longer taboo. It is only a matter of time before the discipline becomes acceptable in broader circles.' In the article, Ricardo Hausmann, Professor of Economic Development at Harvard University, stated that 'Tropical, landlocked nations may never enjoy access to the markets and new technologies, they need to flourish in the global economy.' In his *Physio-economics* (2001, p. 8), Philip Parker went a step further, concluding that distance to the equator is highly predictive (R2 = 70%) of the wealth of a nation – a connection that has proven increasingly powerful over the years. Certainly the statistical connection between latitude and wealth is many times stronger than the one between religion and wealth, as put forward by many other researchers including Max Weber.

However, it is unwise to rely entirely on location and climate factors when seeking explanations for differences in wealth, as it would reek of the old geographical determinism. The link between climate and economy is better seen as indirect than direct. Regions at the same latitude can differ vastly in terms of natural conditions (coastal locations, gulf streams etc.) as well as political and institutional circumstances. In other words, though climate and location undeniably play significant roles in explaining differences, they are by no means all-determining, as we saw earlier from the example of Switzerland.

Box 1.1

Silicon Valley, the world's most innovative region

The name Silicon Valley, first used by American journalist Don Hoefler in the early 1970s, referred to the emerging semiconductor industry in the southern San Francisco Bay area, a region stretching from Palo Alto to the north to San Francisco airport to the south, and from the Santa Cruz Mountains to the southwest to San José to the west. Prior to the arrival of high-tech companies, the region was known for its excellent plum orchards. The valley had a pleasant climate and none of the manic traffic of American metropolises.

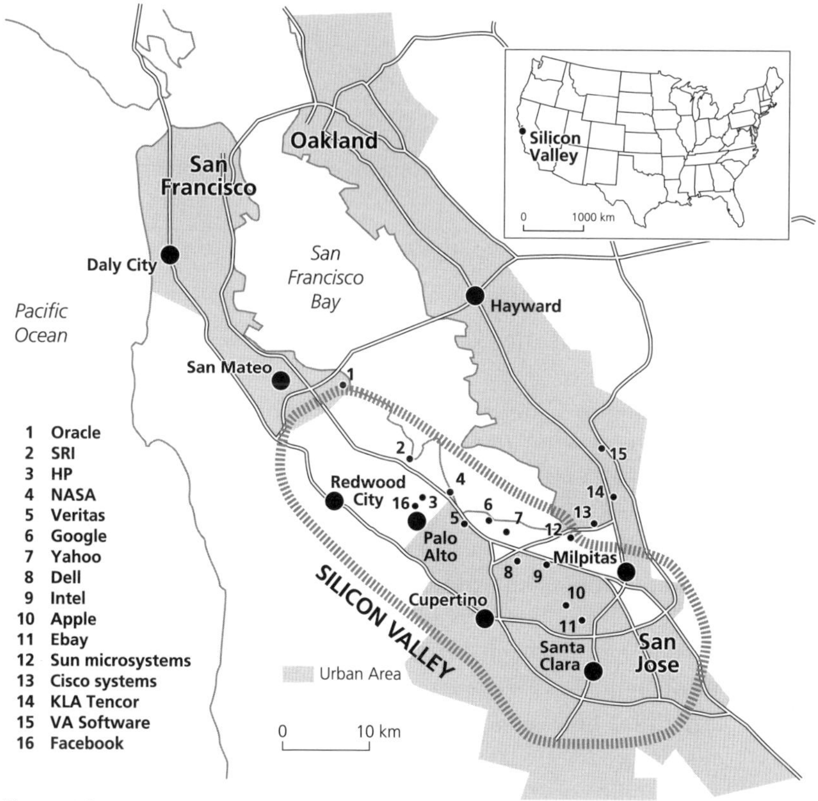

Figure 1.3
Silicon Valley

Silicon Valley, home of the headquarters of Apple, Cisco, Google, Intel and Yahoo, is often considered the nursery for the semiconductor, the personal computer, the Internet and Google. Fredrick Terman, former Dean of the Stanford Engineering Faculty, is generally regarded as the Father of Silicon Valley. His aim was to create jobs for graduates near the university so they would not need to move to the East coast. In 1939, Bill Hewlett and Dave Packard built an oscillator for the Walt Disney

studios in a garage in the Valley. In 1977, local boy Steve Jobs began to build Apple computers in his parents' garage.
Silicon Valley is an example of Schumpeter's 'creative destruction.' Companies based in the area continued to develop. They switched from building hardware to creating software and developing applications and search engines. The innovations that came out of Silicon Valley changed the world and led to dramatic changes in other sectors as well, including the entertainment industry (music, film), newspapers and the publishing world.
Efforts to replicate Silicon Valley elsewhere (Bangalore, Tel Aviv, Côte d'Azur) have so far been unsuccessful. According to Gordon Moore, one of the founders of Intel, a number of factors that led to the success of Silicon Valley cannot easily be replicated:

- A continual influx of highly trained engineers
- An enterprise culture with no fear of failure
- Ample venture capital available
- A government policy creating the right growth conditions.

Source: Markoff, J., 2009

1.2 Location, distance and networks

According to Crevoisier (1999), there are two main opposing views in economic geography, the homogenizing and the particularizing approach. The first includes classical and neoclassical theory, focuses entirely on finding regularity and patterning and ignores historical and spatial deviations from the theory. The second focuses on finding explanations for the particular and unusual and embracing temporal and geographic deviations within the economy. As Crevoisier (1999) noted, economics is shaped by regions, not vice versa. Crevoisier concluded that theories formulated by economists should 'pass through the sieve' of geographers and historians, since history and geography determine how economic theory should be applied in a specific context. Economic geography is effectively area-specific economics. We note above how geographic concepts such as location, distance, density and distribution are once again making inroads in economic science. As the *World Development Report 2009* (*Reshaping Economic Geography*, p. 5) puts it, 'Economic growth is seldom balanced, efforts to spread prematurely will jeopardize progress.' In the view of economists, location is mostly a matter of physical location (costs) and of size and agglomeration (benefits). According to the World Bank, location is of such crucial economic importance that efforts to distribute the economy geographically through policy are futile. This runs counter to the regional economic policy prevalent at the time of the welfare state, with authorities actively attempting to distribute economic growth across their regions by means of growth centres, infrastructure and subsidies. Clear examples are the relocation to the northern Dutch periphery of a number of government institutions and state-run companies, such as the PTT, the forerunner of KPN and PostNL, or to the southernmost region of South Limburg, such as part of Statistics Netherlands and the Algemeen Burgerlijk Pensioenfonds, the largest pension fund in the Netherlands. This policy was directed from the central government in The Hague. Nowadays, spatial geographic economic policy is delegated to local and regional authorities such as provinces and municipalities. The national authorities concentrate entirely on stimulating the top industries in order to reinforce or maintain the international competitiveness of the Dutch economy.

In his 2001 article 'Distance Still Matters. The Hard Reality of Global Expansion,' and his 2011 book *World 3.0. Global Prosperity and How to Achieve It*, Professor of Global Strategy Pankaj Ghemawat drew attention to the role of distance, a factor widely underrated by economists. Contrary to O'Brien and Friedman, Ghemawat did not believe the rise of the Internet had drastically changed the importance of distance. Rather than restricting the scope of the concept to physical distance, he distinguished four dimensions: geographic, cultural, administrative and economic distance. According to his calculations, a 2400 km increase in the geographic distance between two countries will bring trade to an almost complete standstill. Essentially the same point was made by Leamer and Storper (2001), who noted that the Internet hardly affected the geographic clustering of economic activities. Ghemawat stated that cultural and administrative distance have a stronger effect on international trade than geographic distance. For example, companies in countries with a colonial past are ten times more likely to have trading partners in former colonies than in countries without ties of this kind. A common currency boosts trade by 340 per cent and a regional trade block leads to a 330 per cent growth in trade. Ghemawat's conclusion was that if the problems caused by these four types of distance are taken into account, industrial globalization suddenly becomes far less attractive than is generally thought.

Summarizing the force of economic geography, prominent economic geographer Ron Martin said it applies insights from economics, political science, sociology and psychology to what actually takes place in physical space (Martin 1999). In contrast to economists and 'geographic economists,' economic geographers view this physical space as a living space: the actual locations of economic activities and the interaction between them. However, this does not mean current economic geography is restricted to describing the concrete economic reality as it used to be. Like economic science, it focuses on giving explanations, and in doing so it does not refrain from utilizing abstract concepts and theories. An example of this can be found in the article 'Proximity and Innovation' by economic geographer Ron Boschma (2005). He argued that geographic distance is neither a necessary nor a sufficient condition for economic renewal (i.e. innovation), and that this is mainly determined by cognitive and organizational proximity. Though such forms of proximity can potentially replace the effect of geographic proximity, they often go hand in hand. However, Boschma also cautioned against too much proximity, whether geographic, cognitive, organizational, social or institutional, which can be equally harmful to innovation. In this kind of scenario, people are too close to be able to convey fresh news to each other and too hemmed in by bureaucratic regulations. They become too indulgent to each other, lose sight of the economic rationale and as a group suffer from a lack of openness and flexibility.

Maintaining relationships outside the immediate environment can counter these harmful developments. Today's buzzword in this respect is 'networking,' with collaboration between agents as its central concept. They can be companies, institutions or authorities that collaborate to save money, exchange knowledge or inspire each other. For many companies, what the consumer wants is a key stimulus behind the networking activity. Marketing and innovation are increasingly interconnected, and what the modern economic geographer is wondering is how these networks influence the behaviour of manufacturers and consumers. An important question that ensues from this is how these networks can be organized. In this connection, British geographer Peter Taylor, who laid the foundation for the 'world city network approach' (2001), noted that 'as a product of 'state-istics,' the geographical bedrock of social sciences is territorial; it describes the world as a space of places.' To him this was too narrow a view. What is needed is research analysing the world as a space of interactions. Taylor reached this conclusion after studying international networks such as those of banks and solicitors' practices. He argued that it is impossible to explain the location choices of large corporate service providers without insight into the information exchange between these organizations and the various networks they operate in. Here Taylor agreed with Spanish sociologist Manuel Castells (2000), whose name is associated with spatial networks and who was the first to state that 'spaces of flows' may well be more important than 'spaces of places.'

1.3 Industry and environment

The relation between economics and geography has not only given rise to debate within the academic community, it also has a strong bearing on industry. It is in business management that the two disciplines meet. This multidisciplinary area of study is concerned with optimizing managerial processes. It does so by studying how businesses are organized (e.g. in terms of staff, purchase, logistics, marketing and financial management) and the environments they operate in. The paths of economic geography and business management cross at the point of the latter key objective, as both disciplines study the factors determining where a business decides to establish itself.

Location decisions with an international scope are often about access to a foreign market and better protection of the company's own products. So a central location in a continent can play an important role. Once a company or its subsidiary has established itself at a given location, it starts to build regional connections. They can be business links with suppliers, consumers and ancillary services, or links with authorities, for example concerning legal aspects. However, the connections can also be of a more socio-cultural nature, such as sponsoring local clubs and charities or taking on responsibilities in the interest of the sustainable development of the immediate environment.

Each company views the regional environment from its own perspective. To businesses, the environment in the spatial sense of the word appears to be a flexible concept. A small bakery with a regional reach will be less influenced the expansion of the European Union than a multinational like Philips. Changes in EU legislation on the transport of goods may not affect a service provider but will impact an international haulage company. The significance of these types of spatial factors also varies per company. Agricultural businesses are strongly affected by environmental legislation, but the very same rules and regulations are far removed from the world of an optician at a local shopping centre. In other words, not only is the scale relationships are played out on important, so is the nature of the relationships.

However diverse regional factors are, they are increasingly becoming less localized and more globalized. The growth of globalization is demonstrated by a wide range of indicators. For many years, trade has been outgrowing production and international investments are growing even faster. According to Friedman (2005), the symbol of this new age is the Internet. Rather than the amassing of nuclear power that typified the Cold War (often symbolically represented by Albert Einstein's $e = mc^2$), the current age of total globalization is governed by Moore's Law, which says the performance of computer chips doubles every two years while the price is reduced by half. In the Cold War period, everything revolved ideologically around the people who held the reins of capitalism, as Karl Marx and John Maynard Keynes each pointed out in their own words. Friedman's ideas about the present age correspond with those of Austrian economist Joseph Schumpeter, a strong proponent of continuous renewal, with innovation replacing tradition. Or to use a sporting analogy, the Cold War was a wrestling match where everyone circled around everyone else but no one got hurt, whereas globalization is a high-jumping competition. For years, athletes jumped face forward after a straight approach, but when Dick Fosbury introduced the Fosbury Flop, clearing the bar in a supine position after a curved, sideways approach, this instantly made him Olympic champion at the 1968 Mexico Games. Although athletes have continued to perfect the technique, they never looked back and no one jumps the old way any more. Friedman views this kind of 'creative destruction,' a term borrowed from Schumpeter, as the most typical characteristic of a globalizing world.

The trend towards ever-increasing globalization also implies geographic consequences. The Cold War was dominated by strained relations between the United States and the Soviet Union. Today, in addition to the traditional balance in power between nation states, we are dealing with relations between nation states and financial markets. The 2008 financial crisis, known as the Credit Crunch, highlighted the strong interdependence of financial markets globally and the extent to which how banks and other financial institutions act can influence national policies. In the sixth edition of his monumental book *Global Shift*, British geographer Peter Dicken (2011) demonstrated how the ratio between daily financial transactions and actual trade rose from 2:1 in 1973 to 100:1 in 2007. While partly linked to trade and investment, these transactions are increasingly aimed at short-term profit and speculation.

The figures above indicate a shift from regional to global competition. Once emblematic Dutch bicycle brands like Gazelle, Fongers, Sparta and Batavus are now all owned by foreign firms, and the Taiwanese bicycle manufacturer Giant has started manufacturing bikes in the Netherlands. Another example is the flower and plant sector. Dutch flower growers have set up shop in Kenya and are now competing with their colleagues in the Netherlands at the Aalsmeer flower auction. Yet this has not destroyed the Dutch flower sector. Although Dutch flower production has decreased, the Aalsmeer auction continues to be the main hub in the distribution of cut flowers and pot plants. The Dutch auctions have evolved from national sales hubs to international market places. This trend, with directing operational processes and enhancing the trade function playing an ever more important role, is also evident with respect to other agricultural products. Here too, the Dutch have successfully specialized in buying and selling and mediating in trade transactions. Globalization has driven down transaction costs, leading to increased specialization and the fragmentation of production. Naturally this also applies to other countries and regions. In India, for example, software is being developed for Western companies on a large scale, firmly putting places like Bangalore on the map. This kind of international competition has led to a situation where many activities are simply no longer profitable in countries like the Netherlands.

The term 'environment' has a broader definition in the discipline of business management than in economic geography. In business management, the environment comprises everything that happens outside the business and has to be taken into account in its management. It can involve developments on the foreign market or political or legal decisions made at the European Union level. In economic geography, the semantic scope of the concepts 'functional' and 'spatial' is restricted to factors that influence why an economic activity is located at a particular place. In this context, Dutch economic geographer Marc de Smidt (1941-1992) defined the production environment as 'all the external conditions that influence both the decision to locate a business in a particular place and how it subsequently functions' (1975, p. 48). In this definition, we recognize the two aspects that characterize economic geography, i.e. location choice and regional development. In business management, where a business decides to settle is seen as a direct consequence of its internal organization. In economics, this type of decision is viewed as resulting from a costs and benefits perspective. The economic geographer, however, views it from a territorial perspective, with the external conditions examined at different levels – local, regional, national and international. This distinction will be discussed in greater detail in Chapter 2.

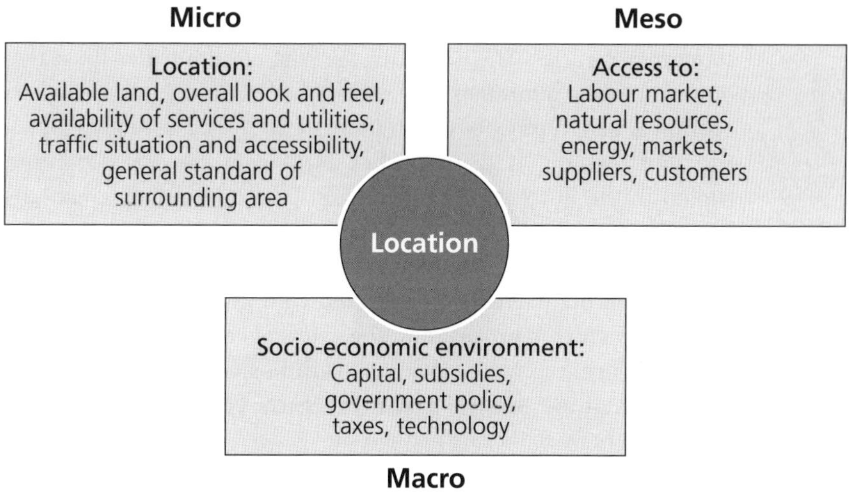

Figure 1.4
Location factors at the micro, meso and macro level

The various regions in the Netherlands have a great deal in common: there is a single tax system, collective agreements ensure virtually equal pay for certain types of labour, all the provinces have the same kind of educational institutions, and all across the country the road network meets the same minimum standard. In other words, for many though by no means all activities, many though again not all the regions are able to provide the basic conditions enabling companies to function.

Nonetheless, the Netherlands does not provide the kind of equal conditions an 'urban field' does. In this kind of region, it does not matter so much exactly where businesses decide to set up shop, since conditions will be favourable to them everywhere. There are local differences within the Netherlands and companies take them into account. This can have to do with the increasing importance of 'soft' location factors, such as the look and feel of the premises, the quality of the surrounding area and the reputation of the region. According to research bureau Stec (2001), the importance of soft factors depends on the type of activity, with sales, consultancy and shared services being the most sensitive to soft location factors. From a comparison of three studies on corporate migration dated 1977, 1988 and 1999, Dutch economic geographer Frank van Oort (2008) concluded that over time, premise representativeness is becoming an increasingly important location factor in the Netherlands.

1.4 Conclusion

In this chapter we have seen how closely related the disciplines of economics and economic geography are; economic geography is the spatial application of economic theories. The desire on the part of authorities for economic development in their administrative regions has led many policy-makers to include economic geographic theory in their toolkit. As a strongly empirical science, economic geography distinguishes itself from classical economics, which is rooted in the development of models. In recent decades, economic science, like other social sciences, has witnessed renewed interest in geographic variation and its influence on economic development. Interdisciplinary research has been on the rise, and by engaging with aspects such as the structuring of the manufacturing process and the external conditions for economic activity, economic geography also overlaps with business management. In Chapter 2 we examine how classical and neoclassical economic theories influence the development of economic geography.

2

Classical and neoclassical location theory

2.1 Introduction

Economic science is about how people cope with the scarcity that forces industries and consumers to make choices. Locations can also be in short supply and not every company can have an optimal location. The choice of location is an economic issue. There are a number of theories on how the underlying process works. In this chapter we discuss the main location theories with a classical and neoclassical underpinning. The basic assumption in the theories is that in an economy where sellers aim to maximize their profits in the face of cost constraints, income-constrained buyers seek ways to maximize their utility. Making rational choices, the theory argues, sellers seek to utilize their factors of production as cost-effectively as possible, and buyers do the same with their disposable income when making purchases. In classical location theory, there is a strong focus on spatial variation in the cost of factors of production and transport. Neoclassical theory also takes spatial variation in profits into account. Rooted in economics, both theories find their main application in economic geography.

A theory is a system of verifiable arguments with a logical coherence. Each argument can be deduced from the previous one. A *location theory* thus consists of a set of connected arguments explaining the choice of location for businesses and services. Although classical and neoclassical theories are historically at the heart of economic geography, there is considerable debate among geographers regarding their validity. In Chapter 1 we note that economic geography is a highly empirical science. In 1969, Professor of Geography Hendrik Keuning (1904-1985) wrote that since it was developed by economists and because of its theoretical basis and deductive nature, there was no place for location theory in an empirical science like geography. Nevertheless, these theories were very popular in the discipline of systematic geography (spatial science), which was on the rise at the time, and classical and neoclassical location theory experienced a revival in 'new economic geography,' which is dominated by economists. Classical and neoclassical location theory may be old but they can still be useful.

2.2 Classical economics: Minimal costs

With *The Wealth of Nations* (1776), Adam Smith (1723-1790) laid the foundation for classical theory in economic science. One of the basic principles of classical economics is that each type of supply creates its own demand. The manufacturer is thus able to use the available factors of production to the full extent and has no difficulty selling the finished products. Classical economics also argues that the availability of factors of production, such as labour, capital and resources, determines the opportunities for industry. Factors of production are paid for in the form of wages, interest, leases, rent or purchase costs. As a result, entrepreneurs are faced with a problem of allocation, i.e. how to achieve maximum production at minimum cost. Since costs vary per location, there is also a geographic element to this question. In addition, the answer to this question is relevant to the wealth of nations, as countries with low average costs are at a competitive advantage.

The main proponents of classical economic location theory were Johann Heinrich von Thünen (1783-1850) and Alfred Weber (1868-1958). Von Thünen focused on agricultural land use, and Weber mainly on industrial location choice. They both held that an entrepreneur would choose the location with the lowest overall costs. We should note in this connection that classical theories are based on the premise of a situation of 'perfect competition' on the markets, i.e. numerous sellers, none of whom can individually influence the price of a product. The price is fixed and competition is based on costs. Geography plays a role, since there is spatial variation in the cost of land, labour and capital as well as the transport of incoming raw materials and semi-finished products and outgoing finished products. The most profitable location is the one where the combined production and transport costs are lowest.

Classical location theories are based on the assumption that people are fully informed and act rationally – the 'economic man' or 'homo economicus' – and the world is seen as an 'isotropic space.' This is a theoretical construct, a uniform space without geographical barriers (mountains, rivers, infrastructure, socio-cultural differences etc.). In addition, classical location theory is deductive. On the basis of logical arguments, patterns are discerned that apply across the board. In other words, it is not the strange or the deviant that are of interest, but the commonalities.

2.2.1 Agricultural land use according to Johann Heinrich von Thünen

Figure 2.1
Johann Heinrich von Thünen
Source: Johann Heinrich von Thünen-Institut, Braunschweig

In 1826, Johann Heinrich von Thünen published his theory on agricultural land use. As a landowner in the German region of Mecklenburg, he observed that land use changed as the distance to the market increased (in his day, markets were always in the nearest village or town), and Von Thünen wondered why this was the case. It was this insight that led him to develop his theory. Rather than focusing on why a farm was located in a particular place, his theory concentrated on what crops were grown. For example, Von Thünen wondered why arable farming was located in different areas than dairy farming and horticulture. Following classical economic thinking, he assumed a farmer's decision to grow a certain combination of crops was a purely rational choice. Farmers, he thought, saw crops in terms of costs and returns.

Box 2.1
Johann Heinrich von Thünen

As a landowner, Johann Heinrich von Thünen (1783-1850) was able to combine theoretical mathematical knowledge with practical experience at the farms run by his tenant farmers. Von Thünen studied agricultural science as well as economics. Via his wealthy wife he acquired a 1150-acre estate near Teterow in Mecklenburg. He used the estate to conduct research on soil fertility and trends in grain prices and agricultural wages. In 1826 he published his seminal work *Der isolierte Staat in Beziehung auf Landwirtschaft und National-Oekonomie (The Isolated State).* Von Thünen created a formula for the calculation of a fair wage – the root of the product of the minimum living standard times the productivity:

$$A = \sqrt{ap}$$

Von Thünen gave his tenants a share in the profits. He put money aside for them and did something very out-of the ordinary in those days: when they turned sixty they received a 'pension.' There is a museum dedicated to his life and work in the German town of Tellow.

Source: www.Thuenen.de, Dievorst *et al.*,1984.

According to Von Thünen, land use derived from the price of land and expensive land (fruit, vegetables) was not cultivated the same way as cheaper land (rye, pasture). Classical economist David Ricardo (1772-1823) had already drawn attention to variations in lease rates caused by varying levels of fertility, i.e. the 'economic rent' of the land. Von Thünen adopted this concept and added the cost of transport to the market (the 'location rent'). These, he believed, were the decisive costs.

The deductive theory used in classical economics works on the basis of assumptions regarding reality. In Von Thünen's theory, the following assumptions apply:

- Space is isotropic: flat and boundless, with no natural barriers. Natural resources and climate are the same everywhere.
- Land is uniform: farmers can have the same animals and grow the same crops anywhere.
- There is one single market: the nearest town or village.
- Transport costs vary per product.
- Transport costs increase linearly as the distance to the market increases.
- There is one single mode of transport: all products are taken to the market by ox-drawn cart.
- The return per hectare is the same for all the crops in the area.

With these assumptions as point of departure, a farmer aiming to maximize his profits (W) had a clear goal in mind. He needed to make sure the difference between a given market price (VM) and the sum of the production costs (P) and transport costs (T) was as large as possible. This translates into the following formula:

$$W = VM - (P + T)$$

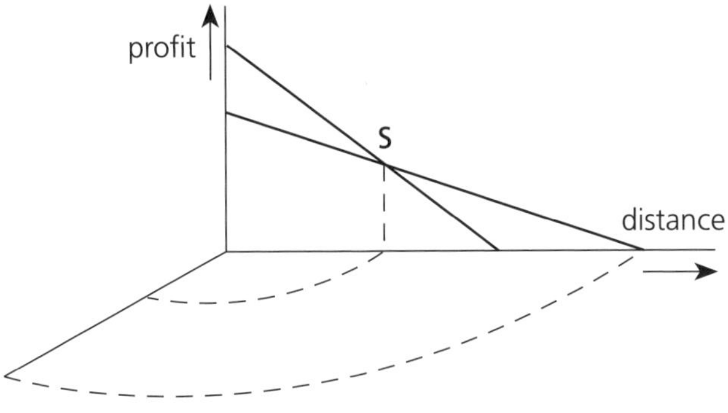

Figure 2.2
The effects of transport costs on the price at the market

Since the competition was perfect, Von Thünen used a fixed market price (VM). Figure 2.3 shows how the overall costs in market M are equal to the production costs. This is because no transport costs are incurred in the production process in market M (goods are produced and sold at the same location). From the perspective of transport costs, farmers in M make the largest profit. Production outside M causes profits to fall, and the decline increases as the distance to M increases, since the farmer then has to pay higher transport costs. As Figure 2.2 shows, in S there are thus no profits to be made with this particular product at all. There the total costs (the sum of the production and transport costs) are equal to the selling price. Outside zone MS, not a single farmer can be found who grows this particular crop.

However, farmers have a choice, and they can grow rye, wheat, beets, cabbage, maize and so forth. A range of options like this generates a different scenario. Based on the economic rent of the land and the cost per product unit of transport to the market, Figure 2.3 shows how the farmer can judge which crops would be best to grow. While it costs less to grow crop A (MA) than crop B (MB), the transport costs for A increase faster per kilometre than those for B. The question Von Thünen tried to answer using this example was: Which crop is grown where? He followed the line of thinking shown in Figure 2.3. The answer is that crop A is only grown in area ML and B in MX. At every location, the farmer has to make choices regarding land use. For area M to be productive, he needs to grow crop A, since the production costs per hectare for A are lowest there and the net return (selling price minus costs) highest. In the area between M and N, farmers grow A as well, since it is more profitable than B. In N, the two crops are equally profitable and beyond N, farmers grow B.

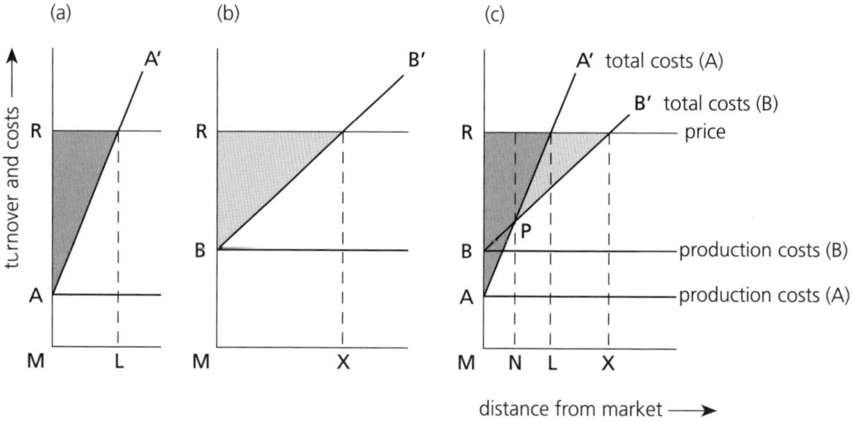

Figure 2.3
Transport and production costs per hectare for two equally profitable crops

In reality, the returns differ per crop. A hectare of tomatoes, for example, is worth more than a hectare of wheat. This leads to yet another scenario (see Figure 2.4).

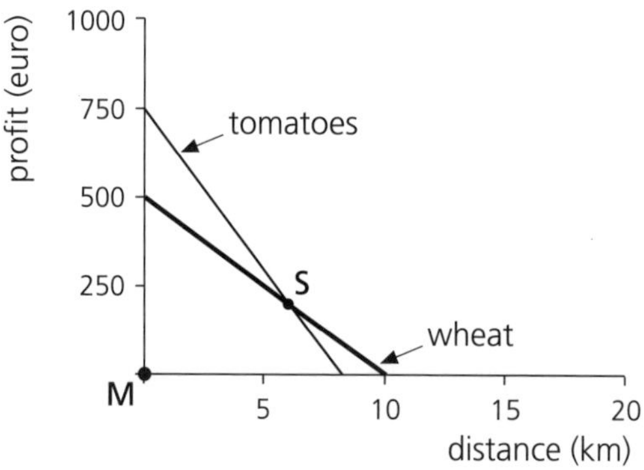

Figure 2.4
Returns, transport costs and production costs for two crops (tomatoes and wheat)

In market M, the farmer will choose to grow tomatoes because the returns minus costs (but not transport costs!) per hectare for tomatoes are 750 euros as opposed to 500 euros for wheat. Outside M, tomatoes rapidly lose this edge in the face of relatively high transport costs. Further than six kilometres from M, the farmer would do better to grow wheat.

	Tomatoes	Wheat
Returns per hectare	€1500	€1000
Production costs per hectare	€750	€500
Transport costs per km*	€100	€50

* For quantity of produce per hectare

The presupposed isotropic space and single market allowed Von Thünen to present his analysis as a three-dimensional figure consisting of concentric circles representing agricultural land use arranged around the market centre (see Figure 2.5).

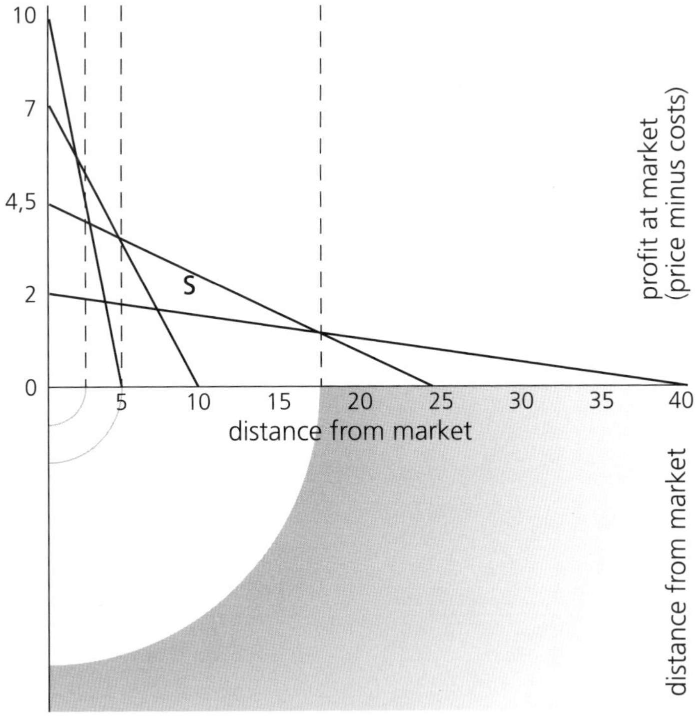

Figure 2.5
The relation between economic rent and land use in concentric zones

Von Thünen was aware of his theory's limitations. In later versions, he considered the spatial variation in climate and soil fertility. He also addressed a dual market scenario. These adjustments led to the partial disappearance of the concentric circles. However, the basic principles of Von Thünen's theory remained unaltered and it is still used today.

Recent research by Sills and Caviglia-Harris into land use in the Amazon region showed that Von Thünen's model still applies today. One of their conclusions was that 'distance to market (…) explains nearly one-third of the variation in farm value, as predicted by the Von Thünen model' (Sills and Caviglia-Harris 2009, p. 66). Another example was provided by regional economist Professor Wim Heijman from the University of Wageningen in the Netherlands, whose 2002 study demonstrated a correlation in Europe between the intensity of land use, population density and agricultural productivity. From 1990 to 2000, according to Heijman, an increase in population density of one person per hectare led to an increase in productivity of $789 per hectare. In other words, there is a positive correlation between population density and the use of artificial fertilizers, which shows that Von Thünen's ideas on the relation between the intensity and the nature of land use and distance to the market are still valid.

In the urban space as well, Von Thünen's model still applies. For example the closer offices are to a city's central business district, the higher the value of commercial property. Not only is the economic rent of the land higher there, there are also location rent advantages for activities involving a high level of interpersonal contact. A central location enhances the accessibility for clients and employees alike (figure 2.6).

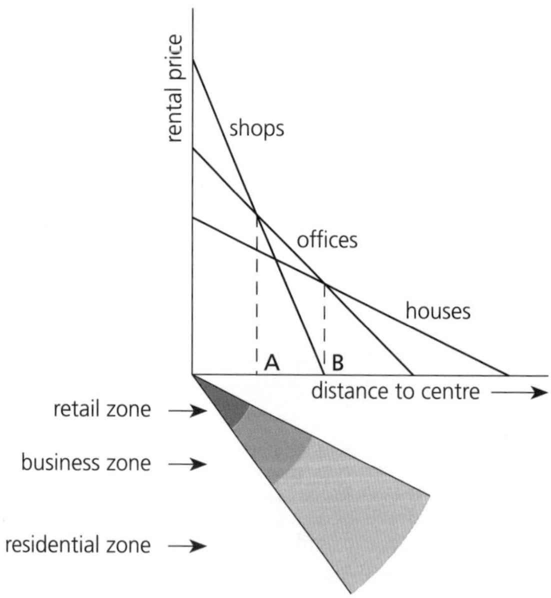

Figure 2.6
Urban location rent

2.2.2 Industrial location according to Alfred Weber

Figure 2.7
Alfred Weber (1868-1958)
Source: University of Heidelberg

In 1909, German economist Alfred Weber published his theory on location choice in primary and secondary industries such as mining and factories. During his lifetime, the Industrial Revolution had led to a high level of industrialization in Germany, mainly concentrated around areas such as the Ruhr district where natural resources like coal and iron ore could be found. The same pattern could be observed in the mining region known as the Black Country in England. In an effort to explain these concentrations, Weber made use of the principles of classical economics. Due to the uneven distribution of the natural resources needed by this industrial sector, the location factors emphasized by Weber were not the same as the ones put forward by Von Thünen. Like Von Thünen, however, he assumed that businesses would aim to keep their transport costs to a minimum. In the case of primary and secondary industries, this not only meant finished products leaving the factories, it also meant raw materials and semi-finished products coming in. In addition, Weber's theory took spatial variation into account in the cost of labour and benefits of scale (agglomeration benefits).

Weber also adopted a deductive line of thinking and used the following set of presuppositions:

- Space is isotropic: the land where companies operate has a uniform topography, climate, technology and economic system.
- Raw materials are unevenly distributed.

- Finished products are sold in a single market place (town or city).
- Though not mobile, as a factor of production 'labour' is freely available everywhere.
- Transport costs for raw materials and finished products are based on the weight of the goods and the distance they are transported.
- Transport costs increase linearly as the distance increases.

Weber termed the factors affecting the choice of location for businesses in the primary and secondary industrial sectors *Standort* factors (the German word *Standort* means location or place of business). He argued that a *Standort* factor had a different value for each location and each business looked for the location with the most favourable combination of *Standort* factors. Like Von Thünen, Weber assumed a situation of perfect competition on the market. So the best location was the one where a business could keep production and transport costs to a minimum. Weber grouped his *Standort* factors on the basis of three criteria:

- *applicability*: general location factors that apply to every type of business, such as transport costs and accessibility – as opposed to specific factors that only apply to certain types of businesses, such as the availability of cooling water;
- *nature*: Weber distinguished two main types of location factors, i.e. natural technical location factors such as infrastructure and societal cultural factors such as regional development support from the central government;
- *functionality*: Weber drew attention to the regional effects of location factors. The availability of natural resources tended to affect a region, since it attracted primary sector industries that processed the natural resources. This in turn drew secondary sector industries, which supplied goods for the regional primary industry sector. He referred to the economies of agglomeration: the cost benefits associated with the spatial concentration of a variety of industries. Benefits of this kind are not only found where natural resources are located, but also for instance in harbours. If blast furnaces cluster together, they can all reap the benefits of their proximity to a deep-sea harbour terminal. In addition to economies of agglomeration, Weber also saw deglomeration as a result of the diseconomies of agglomeration: costs incurred due to the clustering of a range of industrial sectors. One example is traffic congestion.

Initially, Weber's theory was limited to regional location factors: transport and labour costs. Later he also took agglomerating and deglomerating location factors into account.

It is striking that Weber did not include the cost of raw materials in his location theory. He assumed their purchasing costs would be included in the cost of transporting them to the factory processing them so the cost of raw materials was part of the transport costs. Having said that, raw materials played an important role in Weber's location theory. He distinguished two types: 'ubiquitous material' and 'localized material.' Ubiquitous material can be found everywhere. Since the costs do not vary across the region, it is not relevant as a location factor to businesses. They effectively attract the business to the market. An example would be the production of oxygen from air, which is ubiquitous. Oxygen is a vital substance in the chemical industry. Air is freely available everywhere and the transport of oxygen gas is very costly, so oxygen factories are often located in close proximity to their main clients, such as Linde Gas Benelux Ltd in Schiedam in the Netherlands, which supplies oxygen to the petrochemical industries in the port of Rotterdam. Water can also be a ubiquitous commodity. Nearly every village used to have its own brewery. Water was everywhere but the transport of beer, which largely consists of water, was costly. Nowadays there are far fewer breweries, and they are mostly located near reliable sources of good quality water.

The Internet is another context where the term ubiquitous is often used nowadays. Thanks to the World Wide Web, information from all across the globe is now available everywhere. Entrepreneurs whose core business is gathering and distributing information, such as press photographers, tend to set up shop near their main clients, such as publishers. As they can now work from anywhere, photographers are likely to choose to remain in their main markets, i.e. cities like New York or London, where leading publishers such as Reuters can be found.

Where 'localized material' (locally available raw materials) is processed depends on the weight loss that occurs in the process. If industries use *weight-losing raw materials*, they tend to be located near the sources of the raw materials. Crude petroleum was once used to only produce lamp oil, in a process that used a small percentage of the petroleum – the remainder was burnt. Since it was not profitable to transport the oil across large distances and then burn most of it, refineries were located near oil fields. The same applied to the early blast furnaces. A vast amount of coal was required to fire them, most of which was wasted during the production process. There was no point to transporting the coal across large distances, so the first blast furnaces were located near the coal mines, e.g. in northern France.

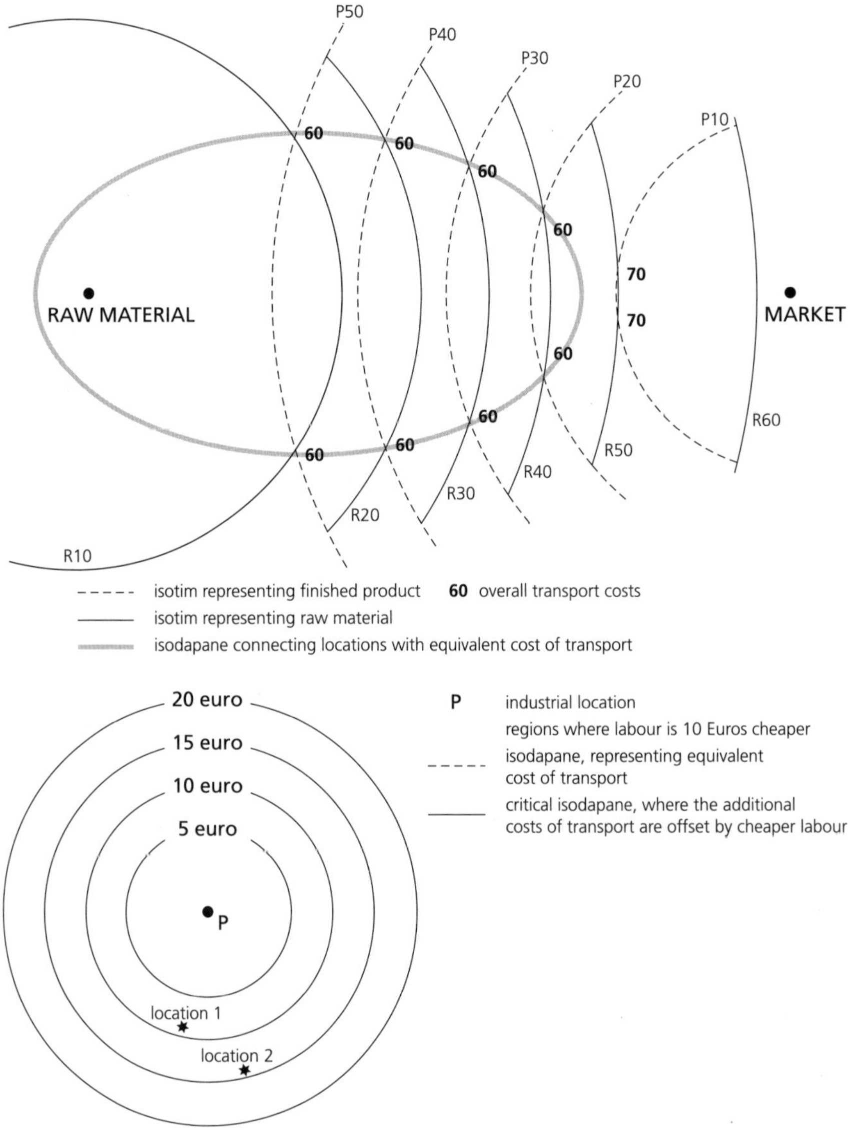

Figure 2.8
Raw materials, transport costs, labour costs and the location of primary and secondary sector industries

If there is little or no weight loss during the processing of raw material (figure 2.8), as is the case for example with mineral water, in terms of transport costs it does not matter much whether the processing facility is near the source of the raw material, near the market or somewhere in between. Massive improvements in the efficiency of production processes have greatly reduced weight loss, so in terms of costs it is no longer necessary for the oil-refining industry to locate plants near the sources of raw materials. The only blast furnace in the Netherlands is not located near the former coal mines in the southern province of Limburg but adjacent to the port of IJmuiden on the west coast near Amsterdam. Here the importance of transport costs outweighs that of the costs of processing the raw material. The same applies to the refineries at Rotterdam's seaport.

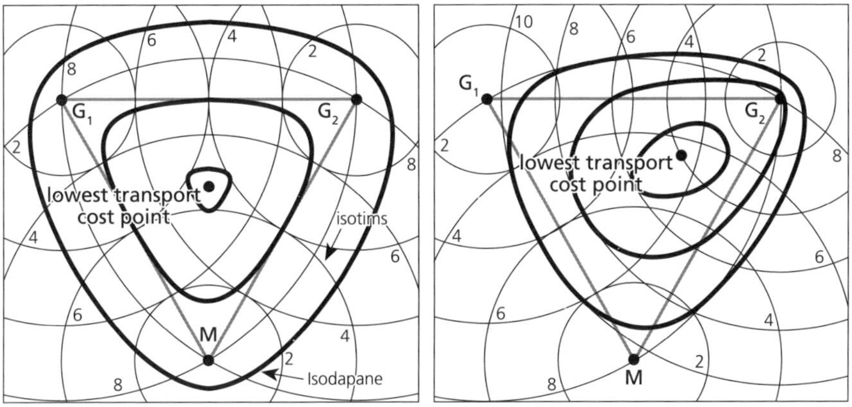

a identical weight loss in raw materials during processing
b varying weight loss in raw materials during processing

Figure 2.9
The relation between location and the minimizing of transport costs: the location triangle

Most industries process a variety of weight-losing materials. Here again, Weber's location theory offers a solution: the *Standortdreieck* or location triangle. Figure 2.9a shows market M and two sources of raw material (G1 and G2). Circles are drawn around G2 connecting the points where the costs of obtaining a particular raw material are identical. Weber called them *isocost lines*. They can also be drawn around the market – lines connecting the points where the transportation costs of the finished product to the market are equal. For each point, the isocost lines can be added up. The points where the overall costs are identical are connected by *isodapanes*, and according to Weber, the point with the lowest overall costs is the ideal location. By overall costs, Weber meant the sum of the total input costs and transportation output costs. In his theory, the transport costs of raw materials and semi-finished products are incorporated into the price when they are delivered to the factory. Figure 2.9b shows what happens when one raw material (G2) loses more weight during processing than another (G1). In such a case, the optimal location shifts in the direction of the material that has lost more weight (G2).

In formulating his location theory, Weber did not just consider raw materials and the cost of transport, he also took the cost of production factors such as labour into account. In Weber's view, if extra transport costs are counterbalanced by cheaper labour, the optimal location is wherever labour is cheapest. Figure 2.10 shows the isodapanes for a given product.

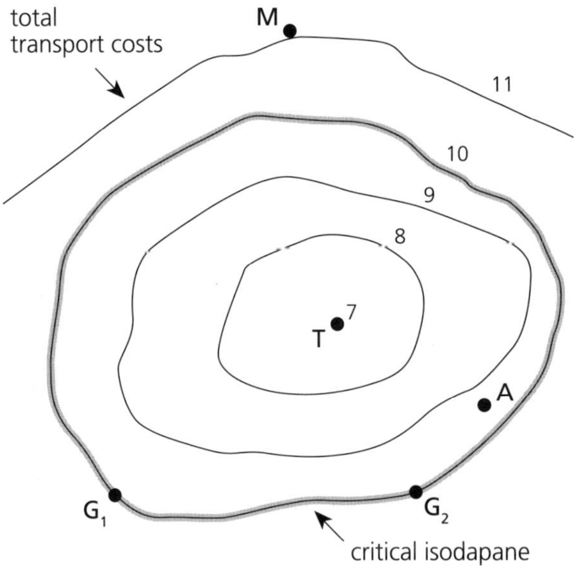

Figure 2.10
Critical isodapanes according to Alfred Weber

Transport costs are lowest within T, namely 7 units. Labour costs are 15 units in all the areas except in A, where they are 12. To determine whether A is a cheaper location than T, the *critical isodapane* has to be found, i.e. the isodapane indicating the difference in labour costs (3) plus the minimum transport costs (7). If A falls within this critical isodapane of 10, as is the case in Figure 2.10, then A is indeed a better location than T. The overall costs in T are 22 (15 + 7), while in A they are 21.5 (12 + 9.5). An example of this type of situation is the location of the first textile factories in the Netherlands. They were not in the cities with large ports, where most of the textile industry's main raw material - cotton - arrived, they were in the southern and eastern provinces of Brabant and Twente. As a result, the cities of Tilburg and Enschede became the main Dutch textile centres.

Like lower labour costs, agglomeration benefits can also prompt industries not to settle at the location with the lowest transport costs. These benefits arise if businesses obtain cost benefits from locations where people and merchants cluster, usually in towns and cities. In the economies of agglomeration, the size and nature of markets are important factors and *localization* and *urbanization benefits* play a role in this connection. Localization benefits are about specialization and businesses operating on the same or similar markets benefit from the proximity of skilled labour and specialist services and the easy exchange of specialist know-how. Urbanization benefits occur wherever variety is the norm and a wide variety of skills and services is on offer for the local business community to tap into. Businesses operating on different markets can also enrich each other via an exchange of unexpected knowledge. Weber's focus was mostly on localization benefits, i.e. the cost effectiveness of labour pooling and collaboration between businesses operating in the same industry. They can be neighbouring merchants that outsource their transport to the same business, which becomes large enough to achieve an efficient cost structure itself. Efficiency gains can then encourage the transport company to lower its rates. Commercial productivity can be enhanced by the opportunities offered by geographic proximity.

Other theorists have since built on Weber's location theory. Two important additions have to do with factor substitution and internal economies of scale. They allow businesses to play strategically with the given costs of external location factors (input and transport costs). Factor substitution means a manufacturer can utilize input goods and individual factors of production in different combinations and ratios, depending on their relative prices. This is an attractive option in situations where relative prices differ per location. At some locations, raw materials are relatively cheap, e.g. at the source, and at others labour may be relatively cheap, e.g. in densely populated developing countries. Altering the raw material/labour ratio in the production process also changes the optimal location. If the use of raw materials increases, the source of the materials becomes an increasingly attractive location, and if more and more labour is needed, businesses will turn to developing countries, where labour is cheap. Stable Western nations have the advantage of a relatively low interest rate on the production factor capital and are consequently favoured locations for the production of capital-intensive goods. The tendency on the part of manufacturers to utilize the factor substitution effect thus changes the relative cost structure of production at certain locations and also of businesses' positions relative to each other.

Internal economies of scale occur if there is a negative correlation between an increase in the scale of production and the production cost per unit. In this type of scenario, increased production leads to relatively lower production costs because of the more efficient use of the available factors of production, lower overheads, lower wholesale prices and so forth. This is why large businesses can produce goods at competitive prices, even if they are not based at the cheapest location in terms of transport costs. Sometimes the location of an industrial facility such as a plant or a factory is highly dependent on the market. An example is the paper industry, discussed in Box 2.2.

Box 2.2

'A cardboard box should never be expensive'

The cardboard market is shrinking because its main clients are relocating to countries where labour is cheaper. Innovation and specialization go some way towards offering a solution. 'Making boxes is a highly critical process.'

One of the greatest nuisances for people who stock shelves at supermarkets are the cardboard boxes most products arrive in. Either they are glued too tightly shut, covered from top to bottom in tape, or they are wrapped too tightly in plastic. A great deal of time is lost trying to open them.

'It could be so much simpler,' says Bert Schoeren, Head of Innovation at Kappa Packaging, the largest paper and cardboard manufacturer in the Netherlands and, with 115 factories, one of the largest in Europe. 'Instead of glue for example you could use a strip to close a box. Or instead of glue you can have a serrated 'press to open' area on the box.' Another way to speed up the shelf stocking process is a box where the front and back can be removed so all its contents can be placed on the shelf in one go. 'With little tricks like that, stocking shelves would go twice a fast,' says Schoeren.

So why aren't the little tricks being used? 'That's because it's not the supermarkets who are our customers, but the manufacturers. They want to see their products packaged as robustly as possible so the contents don't get damaged. They couldn't care less whether a box is also practical in the supermarket.' In addition, packaging equipment is set to the specifications of existing box types. Kappa only aims to develop applications that fit in with the existing production line, so the machines do not require too many adjustments. As Schoeren explains, 'Whatever happens, the box cannot go up in price.'

'Competition is fierce and is based mainly on price,' says Wim Hoebert, Director of Kappa de Zeeuw, corrugated cardboard manufacturers based in Eerbeek in the Dutch province of Gelderland. Hoebert is also Chairman of the Corrugated Cardboard Association, the industry trade group of the sixteen corrugated cardboard manufacturers in the Netherlands who annually produce 1.7 billion cardboard boxes, with a turnover of over 500 million euros. 'Our customers are forever demanding lower prices whereas the high price of paper has driven our purchase costs up, bringing our profit margin down.'

The main buyers of boxes are large manufacturers of food and cosmetics. Together they purchase millions of boxes, giving them a strong negotiating position. They are also highly demanding. 'It may just seem like a simple box,' Hoebert says, 'but actually the production process is highly critical. If a box is not folded perfectly straight, the packaging machine gets stuck and the production line grinds to a halt.'

In recent years the demand for cardboard boxes has shifted from standard boxes with a simple print to multicoloured boxes with the size matching the product. As Hoebert explains, 'A product sells better if it comes in a pretty box, and you can charge more for it. It used to be *the only thing you throw away is the box*, now it's *the only thing you buy is the box.'* Kappa de Zeeuw specializes in multicoloured corrugated cardboard boxes. 'Competition is fiercest in the case of standard products, which in our sector is the brown foldable box. There's a wider margin for specialist products.'

'The art of cardboard making,' says Hoebert, 'is to make sure the production process generates the minimum amount of waste and your stocks are as small as possible.' Corrugated cardboard owes its robustness to the layer or layers of undulating paper, but it takes up a lot of space. 'It mostly consists of air.' Due to its large volume, distribution is costly, restricting the market for corrugated cardboard to a radius of no more than 250 kilometres from the factory. More specialist products have a larger radius. 'Such a limited reach has its advantages,' says Hoebert. 'We are not hampered by competition from Eastern Europe, where cardboard is produced much more cheaply.' However, the recent trend for companies to relocate their production to countries with lower wages has turned this advantage into a disadvantage. 'Part of our customer base is slowly disappearing abroad.'

Hoebert is keenly aware that in the long run, this will threaten the very existence of the Dutch cardboard industry. 'Clearly the market is shrinking. The only Kappa factories where we still see a growth of five per cent or more are the ones in Eastern Europe.' In his view, the sector will have to rely on specialization and the development of new and innovative applications for cardboard, as there is not much more room for increasing profits from enhanced efficiency. 'We already produce twice as many boxes as ten years ago with half the number of people. The production process is highly automated. There's only so much we can do.'

Suppliers in the cardboard industry have also seen their markets shrink. Paper manufacturer SCA De Hoop is located near Kappa de Zeeuw in Eerbeek. It is part of the Swedish firm SCA which, in addition to paper and packaging material, also makes diapers and tissues. In Eerbeek, SCA produces recycled paper for the corrugated cardboard industry and turns used paper into single, double or triple-layered brown paper used for the lining and undulating layers in cardboard boxes, three-quarters of which consist of recycled paper. For the boxes' exterior, new paper tends to be used, providing a better surface for printing.

'SCA de Hoop is highly dependent on the price of used paper, which fluctuates whenever supply and demand fluctuate,' says Head of Production Hennie Harmsen. In his opinion there is a slight overproduction of recycled paper on the European market. 'As a result, prices are low. Sometimes when we have too much we ship it off to Asia, where the demand for paper is high. It's better than dumping it on the market here.' In addition, transport costs are low. 'Many cargo ships bringing goods over from the Far East would otherwise just sail back empty.'

Source: *NRC Handelsblad* 2003

2.2.3 Refining the linear connection between distance and transport costs

Minimizing the cost of transport is a key concept in classical location theory. It assumes there is a linear correlation between distance and transport costs. This is expressed in the formula T = b + aD. T stands for transport and b for production costs as well as loading and offloading, D for distance and a for the product-specific angle of inclination of the increase in transport costs per distance category (e.g. kilometres). This theoretical relationship is shown in Figure 2.11.

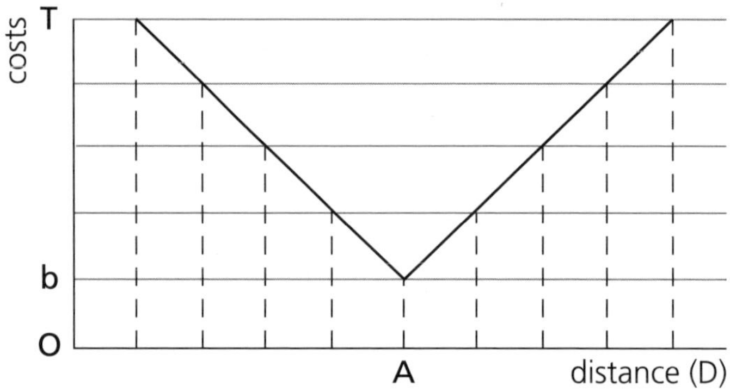

Figure 2.11
The linear correlation between distance and transport costs

However, a linear correlation between distance and transport costs cannot be automatically assumed. The first qualification concerns how businesses calculate the cost of shipping. Classical theories use the 'free on board' or FOB system, in which all the costs of transport are incurred by the buyer and the price includes all the costs from the moment the goods leave the factory or the dealership. This would mean a buyer located near the seller pays less than one based further away. In reality, however, sellers often charge customers an all-in price. This is known as the 'cost insurance freight' or CIF system, in which uniform prices apply within the same market. All the customers at the main Dutch supermarket chain Albert Heijn pay the same price for a bottle of detergent, regardless of how close or far from a distribution centre they live, and gas tariffs are the same in the south of the country as in the north, where the gas is extracted. The cost of transport is no longer a location factor in areas where prices are the same across the board.

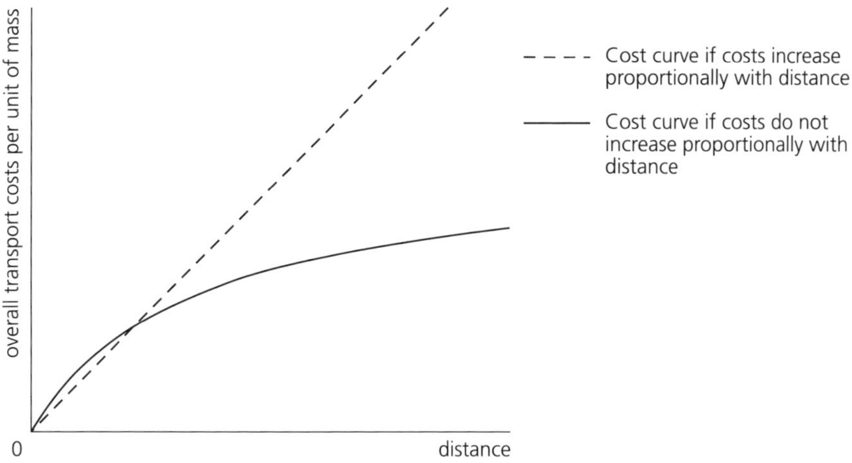

Figure 2.12
The non-linear correlation between distance and transport costs

A second qualification is that the theory does not take into account the structure of transportation costs, i.e. fixed as well as variable costs. Fixed costs pertain for example to loading, unloading and transferring goods and administrative procedures. They are referred to as *terminal* or handling costs, and are not related to the distance. Loading a container onto a railway carriage is not any cheaper if the distance the container is transported is shorter. Administrative costs are not affected by how far goods are transported, although for goods with a long way to go, the fixed costs can be spread across the distance since the fixed costs per kilometre are lower.

Box. 2.3
How containerization changed the role of transport costs

In 1955, American truck driver Malcolm McLean sent a shock wave through the world of shipping. During the long hours spent waiting for goods to be transferred from ships and trucks onto racks and into crates and storage areas, it occurred to him there must be a better way. He proceeded to develop a system that combined detachable trailers and uniform containers, rendering the manual transfer of goods unnecessary. The customers were enthusiastic but the trade unions were not, fearing massive job losses among longshoremen. Many port authorities saw an opportunity to strengthen their position on the market. They worked hard to adjust the ports to the requirements of containerization, the advantages of which eventually brought the unions on board. 'They reasoned that it was better to have fewer workers in a prosperous enterprise than many in a declining one.'
According to the investment bank Morgan Stanley, some 100 million containers were shipped worldwide in 2010 and the bank's forecast for 2020 is close to 300

million. Between the cost effectiveness of containerization and low production costs in Asia, more and more businesses are open to the benefits of moving production to Asia. In 2002, toy giant Mattel closed its last manufacturing facility in the United States. Barbie dolls are now all made in China using Taiwanese plastic, Japanese hair and American colouring agents. Once they leave the Chinese factories, the dolls travel all across the globe. In 2008, the cost of transporting a standard 20-feet (approx 6 metre) container from Shanghai to Rotterdam was 1350 euros. A container can hold 12,500 Barbie dolls, making the transport cost per doll a mere 10.8 euro cents. In 2008, about ten million containers were out at sea every day. Another consequence of containerization is that it reduces the businesses' need to have large amounts of goods in stock.
Containerization has weakened the link between distance and transport costs. Much of the cost of containerized transport pertains to fixed costs unrelated to distance, e.g. costs of handling, Customs clearance and administration.
Sources: Logistiek.nl, Wikipedia, *Nieuwsblad van het Noorden, The New York Times*

The third qualification is connected to the costs of transhipment. The fixed/variable cost ratio is different for each mode of transport. In the case of haulage, where fixed costs are relatively low and variable costs relatively high, it almost amounts to a linear relation between costs and distance. Here too, transport costs level out as the distance increases. Rail transport, however, has high fixed costs and relatively low variable costs, resulting in a stronger levelling out. Figure 2.13 shows that the cheapest option for short-distance transport is road haulage. For longer distances, trains or ships are cheaper. Modes of transport are often integrated to limit terminal costs. Containers arriving at the port of Rotterdam by ship are transferred for road, rail or river transport at *multimodal terminals*. Once in the port's hinterland, the containers are transferred to vehicles that deliver the goods to the customer.

The fourth qualification concerns the relation between the freight rate and the nature of the goods transported. For light bulk goods that take up a lot of space, the transport costs per product unit are very high. The amount of goods also plays a role. For large consignments there are economies of scale, since terminal costs can be spread out across a larger number of goods. Some fragile or perishable goods require extra facilities, driving up the costs. Another important factor is the value of the goods to be transported. The transport costs are less important for valuable goods than cheaper products, and a transport provider can charge more for valuable products.

The fifth qualification pertains to the law of supply and demand for certain trade routes. The capacity requirements for the Shanghai-Rotterdam trade route are much higher than for the Rotterdam-Shanghai route, since far more products are exported from China to Europe than vice versa. So the cost price

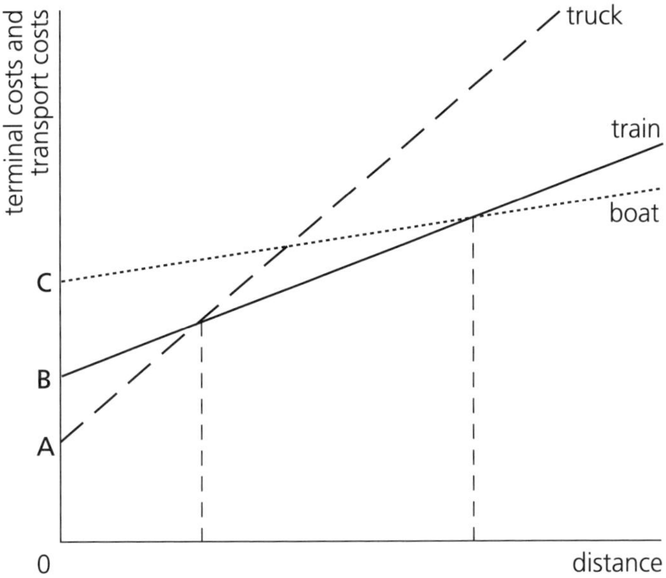

Figure 2.13
The relation between distance and transport costs for a range of transport modes

of containerized transport from China is often much higher (see Figure 2.14). This often makes it worthwhile to fill containers with used paper for the return voyage to China, which would not be possible if the trading relationship was more balanced (see Box 2.3).

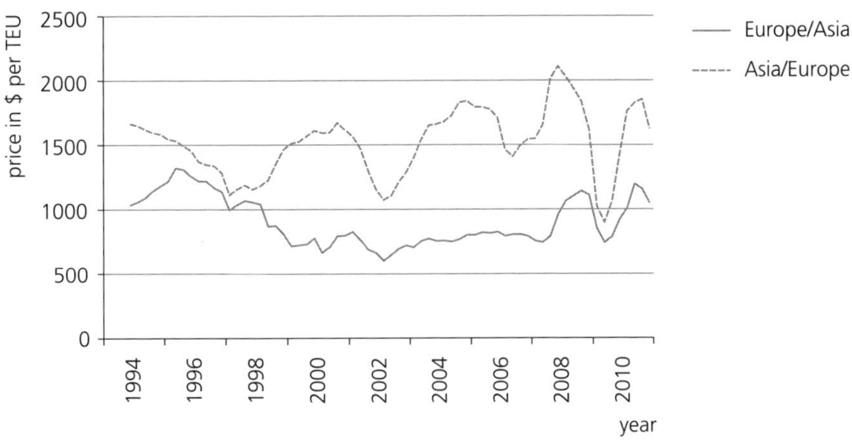

Figure 2.14
Port-to-port container shipment rates in $ for the main transoceanic routes
Source: Dynamar Ltd, Alkmaar, the Netherlands

The five qualifications influence the role played by transport costs in businesses' choice of location. Figure 2.12 demonstrates how transport costs can even decrease as the distance increases. According to Von Thünen's theory, this would mean the area where the incomings outweigh the outgoings becomes larger. Take two competitors who manufacture the same product and are based at locations A and B. If there were a linear correlation between distance and transport costs, the regional market would be divided neatly in two halves. Suppose the company at location A switches to a cheaper means of transport and B does not respond. A is then cheaper and able to expand its market at the expense of B. This example demonstrates that there is more to making full use of a location than minimizing operational and transport costs. In their assessment of the pros and cons of a location, businesses also take into account what their competitors are up to. Insights of this kind mark the transition from classical to neoclassical location theories.

2.3 Neoclassical economics and role of the market

In economics, a distinction is drawn between classical and neoclassical theories. There are five main differences between the two:

- Neoclassical economics devotes more attention to the role of the market. In classical economics the market is no more than a venue for buying and selling. Neoclassical theorists also wonder who controls the market and how businesses can expand their markets.
- Neoclassical theories recognize the existence of more market types than just perfect competition. It is not enough for manufacturers and service providers to look at differences in costs, they also need to devote attention to the spatial behaviour of competitors.
- In contrast to classical approaches, neoclassical theories take into account internal *economies of scale*, highlighting the relevance of the internal organization of a company.
- There are more production functions in classical theories than in classical economics. Neoclassical theories recognize that businesses can create output in a number of ways by utilizing varying combinations of factors of production. Needless to say, the most profitable combination is the preferred one. Neoclassical location theory not only examines the costs of a particular location but also the benefits it can bring, and the assumption is that entrepreneurs are adept at adjusting their use of location factors to their availability. If 'labour' is expensive in a particular area, neoclassical theory says a business will opt for automation, i.e. an increase in the use of 'capital' to reduce the importance of labour to the firm.

2.3.1 Hierarchy and distribution of services according to Walter Christaller

Figure 2.15
Walter Christaller (1893-1969)

German economist and geographer Walter Christaller (1893-1969) investigated the relation between the spatial patterning of rural and urban settlements and their role as regional service providers. He then developed his *central place theory*, also the subject of his thesis published in 1933. Christaller's aim was to formulate universal laws to explain the regularity of settlement patterns. In economic geography, his central place theory is still frequently cited in determining the optimal location for a variety of services, as in the case of the large area of reclaimed land known as the Northeast Polder in the Netherlands.

Christaller discerned a hierarchy of settlements with highly specialist services located in large settlements and providers of basic day-to-day services in smaller ones. The centrality of a settlement thus depends on its size. As regards the role of towns and cities as markets, rather than focusing on transport from the manufacturer to the urban market, Christaller was more interested in the *accessibility* of manufacturers or service providers to urban consumers. He tried to explain settlement patterns (size, number and distribution) in terms of how and to what extent they provide regional services and found that the central-places pattern was the result of careful and deliberate location choices by consumer-focused service providers. For these merchants, more consumers can mean a higher turnover.

Central place theory is founded on a basic pattern every traveller is familiar with. After a string of villages with limited amenities, there is usually a larger settlement with more shops. As towns grow, their function as service providers grows, as does the area they cater to. Not only do urban centres have more shops, they have more specialist shops. Or, as Christaller would put it, they are higher up in the urban hierarchy. In this hierarchy of central places, we can recognize some familiar presuppositions:

- Space is flat and isotropic, so moving goods and people is equally easy in all directions.
- As the distance increases, the transport costs go up.
- Outside the settlements, the population is distributed evenly.
- Income and consumer preferences are the same for everyone.
- Providers (manufacturers or service providers) and users (consumers) are fully aware of what the market has to offer – they are all 'economic men.'
- The market is characterized by perfect competition.
- There are no economies of scale and/or agglomeration benefits.

According to Christaller, a minimum number of consumers, i.e. level of demand, is required for a service or product to be viable, let alone profitable. This *threshold* varies per service or product. The threshold is higher for a supermarket than a pub, which explains why there are so few villages without a pub and so many without a supermarket. At the same time, many villagers would like to have a supermarket. The answer can be to reorganize local shopping facilities. Residents of the southern Dutch village of Sterksel organized a cooperative to reach the threshold for a local supermarket. The village, with 1200 residents, managed to keep a small supermarket operational with one paid employee and fifty volunteers. Christaller's theory says people will always go to the nearest pub or supermarket. The number of customers a local service can hope to attract depends on the maximum distance customers are prepared to travel – the *range* of a service. So the owners of the supermarket in Sterksel not only depend on the loyalty of customers in the village itself, but also of those in the surrounding area.

Box 2.4
Walter Christaller

As a child, Walter Christaller (1893-1969) liked to pore over his parents' atlas. In 1933, after studying economics, he got a PhD in geography. The subtitle of his thesis *The Central Places in Southern Germany* was 'An economic-geographic investigation into the laws governing the distribution and development of urban settlements.'
In 1940, Christaller joined Hitler's National Socialist party, the NSDAP and from 1940 to 1945 he worked at the SS Planning and Soil Office. Part of the Waffen

SS, this department aimed to provide a scientific legitimization for German war efforts to conquer Central and Eastern Europe. In the hands of the Nazis, Christaller's central place theory became an instrument for the planning models for parts of Poland, Czechoslovakia and Russia. Due to his work for the Nazis, his work was shunned by the academic world.
From 1950 onwards, Christaller's work became better known and recognition steadily grew for his central place theory outside Germany. Geographers with a focus on quantitative methods appreciated his work and in 1964 Christaller received the Outstanding Achievement Award from the Association of American Geographers. Towards the end of his life, recognition also came in Germany and in 1968 he received an Honorary Doctorate from the University of Bochum. In 1996 the German Association for Applied Geography named a prize for applied geographical research after him.

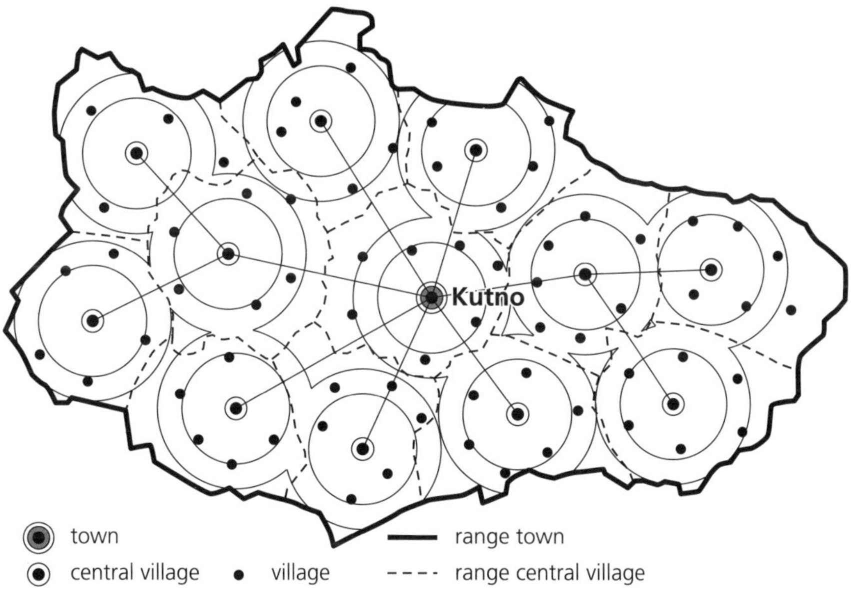

Figure 2.16
Central place theory as applied to Poland for planning purposes during the Nazi occupation
Source: Berry & Chauncey 1970

For all types of services, the market is determined by threshold and range. Based on minimizing the distance within the *isotropic space*, in Christaller's theory these two factors can be viewed as concentric circles around the service. Since service providers cannot be certain all the potential buyers within range will actually use their services, circles representing thresholds tend to be smaller than those indicating range. If not enough buyers can be found within a certain range to make the threshold, the service is not viable. Conversely, if the range far exceeds the threshold, there is room for new providers. Christaller predicted that in this kind of situation, new providers would immediately appear (economic man). In due course, a new balance is reached where the difference between range and threshold is minimal and the overall regional market is distributed evenly across the providers.

Figure 2.17 shows seven market places (M), all with the same characteristics. Each has a service with the range of approximately one kilometre typical of a pub or café. Provided there is sufficient buying power in the market, new central places (M2 – M7) can emerge for pubs beyond distance 'a' of place M1. If the buying power is too weak to keep seven services afloat, the markets (radius a) will grow spatially and the services will have a lower-resolution distribution pattern.

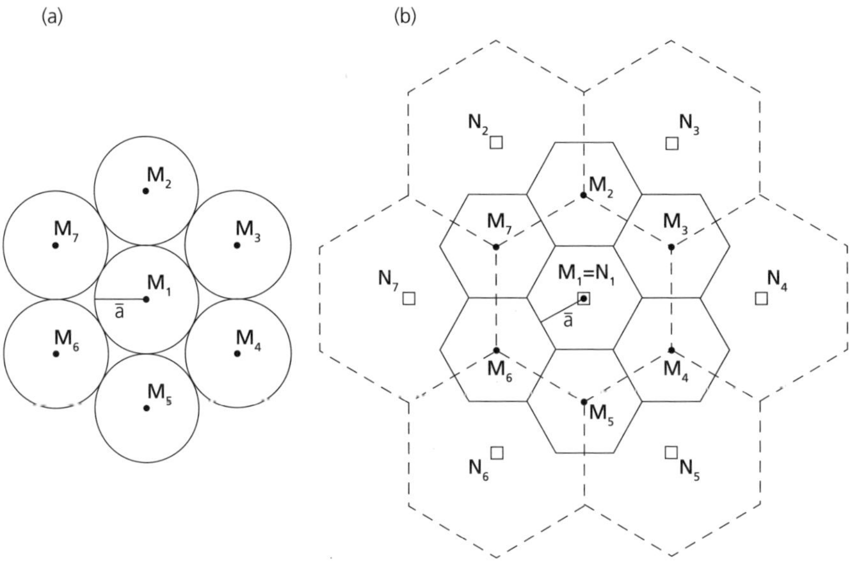

Figure 2.17
Markets in Christaller's theory

Figure 2.17 also demonstrates that if markets are represented as circles, 'empty' areas remain. Christaller assumed the population was distributed evenly beyond the settlements, and people living in the 'empty' areas would not be provided for. In reality however, providers cater to these customers as well. Rather than as circles, markets can more usefully be represented as hexagons, thus covering the entire area (Figure 2.17b).

So far we have only looked at one type of service, the pub. Central place theory really comes into its own if different types of services are considered simultaneously. The threshold for a supermarket is higher than for a pub; the market for a supermarket is greater. Suppose its market is at least as large as the distance between M1 and M2 (Figure 2.17a). In this scenario, the supermarket would base itself in M1 and from there serve customers in the markets in and around M2 to M7. All the places within that area would have a pub but only M1 would have a supermarket. As Christaller put it, M1 is higher in the settlement hierarchy than M2 – M7 since it provides more services, i.e. a pub and a supermarket. Naturally, there are also places with supermarkets further away from M1. These places are represented in Figure 2.17b as N1 (=M1) – N7. This process can be repeated for more specialist services with an even wider range, such as theatres or professional soccer clubs (see Box 2.5). In Christaller 's theory, this process creates a regular pattern in terms of the numbers and distribution of central places. Consumers in M2 to M7 have a choice of three central places, all high in the settlement hierarchy and all equidistant. There is an inversely proportional relation between the number of places and their ranking in the settlement hierarchy (centrality).

Box 2.5
Range and thresholds in Dutch professional soccer

Most professional soccer clubs began as local neighbourhood clubs or initiatives by groups of friends. With the introduction of professionalism in 1954, soccer gained economic as well as social value, and the sport soon came to be run like a business. As with other consumer-focused services, Christaller's theory is useful for examining the location patterns of soccer clubs. To be viable, a club needs a minimum audience, but there is a limit to the distance spectators are willing to travel to the stadium. Soccer lovers will travel further for a Champion League final than a home game of a team at the bottom of the First Division (the second highest soccer division in the Netherlands). For Ajax home games, supporters come from across the Netherlands, but HFC Haarlem, which disappeared from professional soccer in 2009, relied mostly on Haarlem itself and nearby Bennebroek for its supporters. The threshold expressed in numbers of spectators is reflected in the budget. First Division annual budgets range from about one to nine million euros, and teams in the Honorary Division (the highest Dutch league level) range from six to seventy million euros. More spectators mean a higher income, and a higher income means higher salaries, better players and a greater chance of winning. Good soccer results boost the number of spectators. There is a situation of reciprocal relations, with the size of the audience as the basis for a club's athletic and economic possibilities. About a third of a club's income has to come from ticket sales, and the income from sponsoring (40%) and television licenses (20%) is increasingly important. Here too, however, there is a relation between income and potential support base.

Figure 2.18 summarizes the markets of the larger clubs (b – d) and professional soccer attendance (a). Attendance is extremely low in the province of Zeeland, on the Wadden Islands and in the Dutch Bible Belt, a region traditionally home to conservative Protestants. The soccer landscape is not isotropic. In 2008, the level of penetration in the market regions ranged from 109.8 per 10,000 inhabitants in the market of ADO The Hague to 724.3 per 10,000 inhabitants in the market of FC Volendam. For some clubs, high consumption levels can compensate for a small market region.

Source: Dejonghe *et al.* 2006

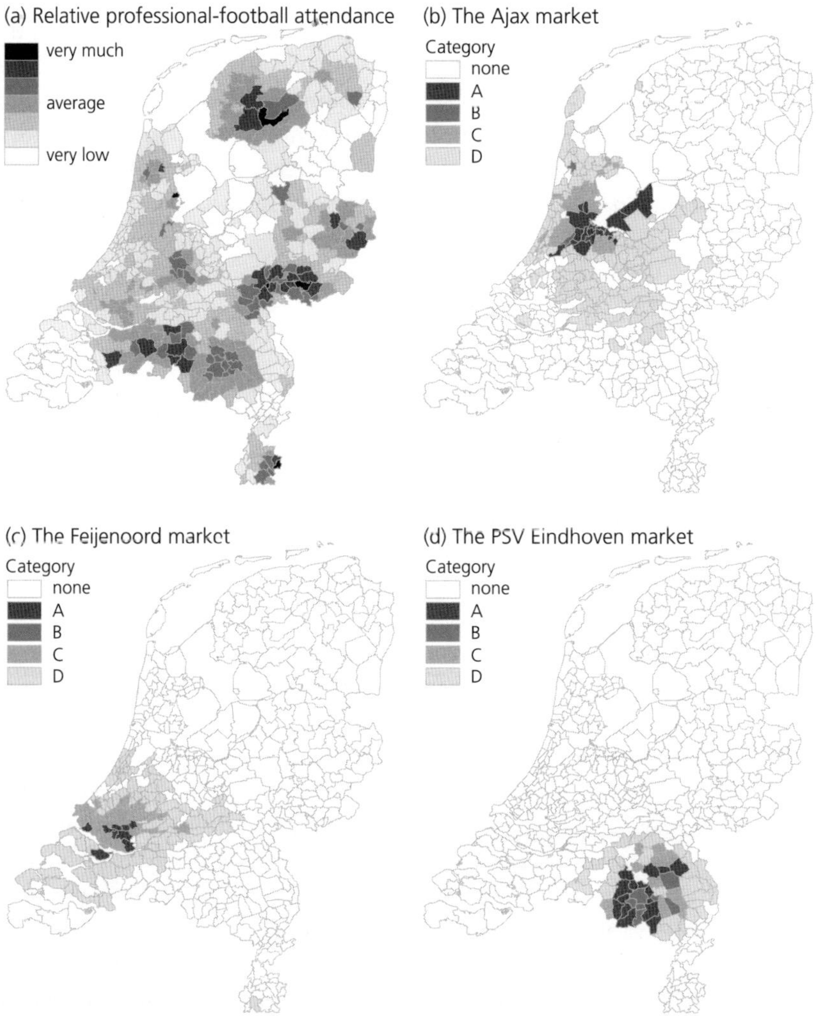

Figure 2.18

A: Relative professional soccer attendance, B: Ajax market, C: Feijenoord market, D: PSV Eindhoven market

Source: Dejonghe *et al.*, 2006

Christaller's theory tells us the location of consumer-oriented service providers is linked to the extent to which they provide specialist services. Diamond merchants are more likely to set up shop in Amsterdam than in a small village, as are exclusive night clubs and high-profile solicitors. This can result from the desire on the part of these providers to monopolize their markets. As is noted above, the theory states that the hexagonal markets are divided up between the providers.

A theoretical link can be drawn between central functions in larger urban settlements and the concept of economic rent as used by Von Thünen. This was noted by William Alonso (1933-1999) using *bid rent theory*, which holds that central places are battlegrounds with providers all vying for the most accessible location. They try to outbid each other, and the highest bidders get the most popular locations. These are the locations with the most economic rent; the highest rents per square metre are paid in the centres of centrally located towns and cities. They are called A1 locations in the Dutch retail sector, and it is here that popular retailers such as large department stores can be found. Not only are they able to pay the high rents, they attract more footfall, enhancing the stream of purchases as well as the turnover for smaller nearby retailers, thus driving rental prices up even further.

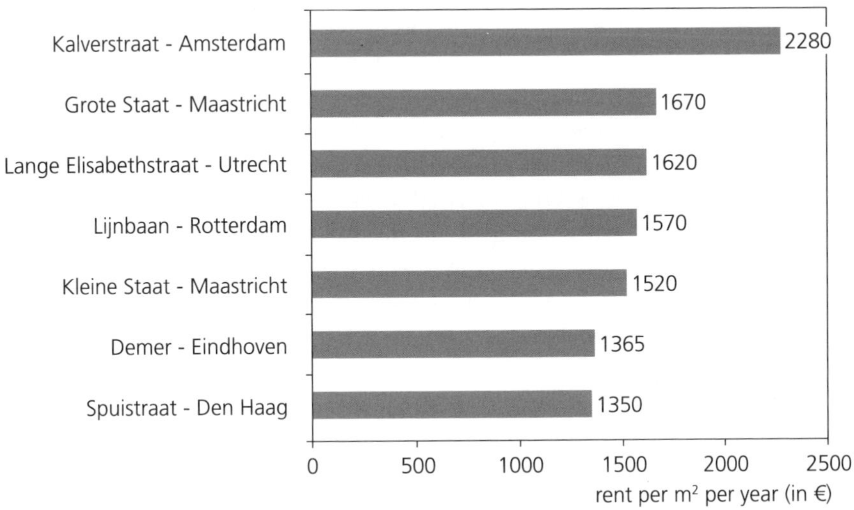

Figure 2.19

Highest annual rental rates for retail property at Dutch A1 locations in 2010

Source: PropertyNL 2011

Popular retailers and product affinity served as the basis for Nelson's 1958 *theory of cumulative attraction*. He believed consumers engage in 'comparison shopping,' comparing retailers on the basis of aspects like the quality, design and price of the merchandise. The clustering of retailers in a restricted area such as a city centre, Nelson stated, offers advantages to buyers and sellers alike.

These theoretical additions are not taken into account in Christaller's central place theory, although efforts have been made to test his theory empirically. A good example is Dutch geographer Jan Buursink in his 1971 thesis *Centrality and Hierarchy*. In another Dutch example, Christaller's theory is applied to settlement planning in the Northeast Polder, a large area of reclaimed land, and it is frequently referred to in the hierarchy of shopping centres from city centres to local neighbourhoods. In determining the composition of shopping centres in terms of shops and other services, developers and managers of retail locations look at the location of their centre in the hierarchy within urban regions. Not every shop or service will thrive in a given shopping centre. The individual retailer will adapt his product range to the market and avoid stocking unprofitable merchandise.

Since some of his assumptions are not confirmed by reality, there are however major drawbacks to empirical analyses and planning applications of Christaller's theory. It is questionable for example whether the principle of the concentration of central functions in central places and their surrounding area applies. Many services have a much wider range than the area immediately surrounding an urban settlement. The urban concentration of services can also result from agglomeration benefits, as is illustrated by Nelson's theory. Furthermore, Christaller's theory presupposes that everyone has the same income and preferences, but in reality there can be enormous differences in purchasing power and taste. This is why market researchers divide consumers into groups based on income level and social class. Another of Christaller's unconfirmed assumptions pertains to linear transport costs (see Section 2.2.3). Shopping centres compete on the basis of accessibility and parking charges. Some try to enhance their distinctive capability by offering a thematic range of services, thus enlarging their markets. The regular hierarchy of central places with associated services that Christaller envisaged often evaporates in the face of reality.

2.3.2 Locational competition according to Harold Hotelling

Mathematical economist Harold Hotelling (1895-1973) is a typical proponent of neoclassical economics. The economic law named after him, Hotelling's Law, states that in competitive markets, providers tend to offer similar products, and in oligopolistic markets, they aim to be distinctive. In other words, the greater the competition the more homogeneous the products, since competitors keep a watchful eye on each other. If one competitor changes a product, others immediately copy the change. Hotelling's theories were adopted in geography because he saw the choice of location as a variable dependent on providers' competitive behaviour. In deciding where to set up shop, they not only consider costs and consumer distribution but also what the competition is up to. They may even try to anticipate the others' next move. This locational connectedness of merchants is sometimes called 'locational interdependence,' and neither classical economics nor Christaller's theory allow for it. Another difference between Christaller and Hotelling is that their theories are based on different market situations, Christaller's on perfect competition and Hotelling's on a duopolistic market, i.e. containing two providers.

Hotelling published his ideas on this last aspect in his article 'Stability of Competition' in the renowned *Economic Journal*. It appeared in 1929, the year of the Wall Street Crash. Hotelling criticized classical economics' 'invisible hand,' which was supposed to ensure economic recovery. His criticism was mainly aimed at the idea that in a situation of perfect competition, each provider specialized in a particular product or target group. Hotelling envisaged exactly the opposite scenario, and held that the producers' aim to maximize profits combined with the consumers' aim to minimize costs would lead to one uniform product designed to satisfy the largest possible consumer group. While taking the need for consumer accessibility into account, businesses' tendency to keep a watchful eye on each other would lead them to set up shop near their main competitor rather than close to their consumer base. Businesses play a 'location game' with each other and with the consumers, each anticipating the behaviour of the other players. Each business tries to obtain the best location, but ultimately they all find themselves in close proximity. In economic geography this is called 'agglomeration.' Unlike the agglomeration factors cited when we discussed Weber (section 2.2.2), the 'locational interdependence principle' is not about reducing costs but about obtaining a favourable market position.

In economic geography, Hotelling's ideas have been elaborated upon using the example of two ice cream vendors arriving at a beach (see Figure 2.20). The beach is completely closed off on both sides. They shake hands and each decides to sell at least as much ice cream as the other that day, and preferably more. They survey the beach to see where their customers, the bathers, are located and conclude that they are distributed evenly across the beach. The obvious approach would be to divide the beach in two halves. This would be in line with Christaller's theory and each would set up shop in the centre of his half (phase 1). However, each wants to sell more ice cream than the other, and after some thought, one of them comes up with the cunning plan of moving slightly towards the middle of the beach. He can still sell to the bathers on his half, but now he can also steal some of the customers on his competitor's half. The most extreme form of this scenario would be for him to position himself next to the other vendor, while remaining within his own half of the beach (phase 2). Phase 1 is unstable. Each vendor has to be prepared for the other's opportunistic behaviour. Only by effectively 'leaping over' vendor A can vendor B avoid a loss of market. If this behaviour is repeated often enough, both vendors would end up in the middle of the beach (phase 3). Only then would equilibrium be achieved, or 'stability in competition.' The losers in this scenario are the consumers, who need to walk further for their ice cream. Whether they do so, as we saw when we looked at Christaller's theory, depends on the range of the ice cream. Essentially, what Hotelling was trying to convey is that competition initially leads to product uniformity and subsequently to the spatial clustering of product providers.

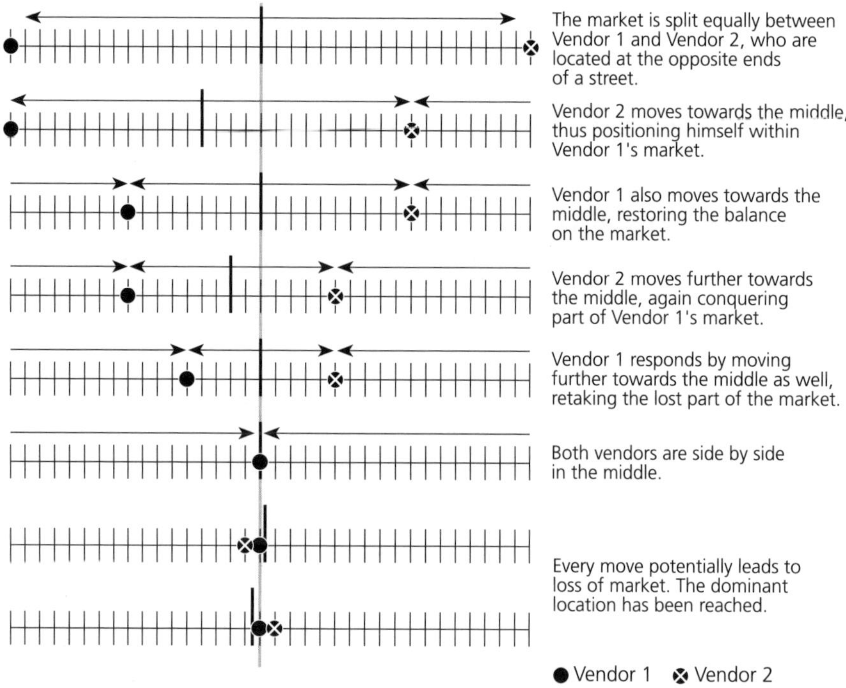

Figure 2.20

Examples of the processes at work in locational interdependence

Source: Sheldon 2011

The ice cream vendors' choice of location in Hotelling's model contrasts sharply with where they would have positioned themselves according to Christaller. He would expect them to distribute themselves evenly instead of clustering in the middle of the beach. He describes a situation of monopolistic competition where the kind of opportunistic behaviour we see with Hotelling is not profitable. Hotelling's duopolistic market is one of income maximization leading to locational interdependence, as illustrated by the example of the ice cream vendors.

2.3.3 Markets according to August Lösch

August Lösch (1906-1945) was a German economist who worked at the University of Kiel in Germany as well as at American universities. A true academic, he considered the actual location of a business less important than the theory behind the best location for a business. In 1939 he published his ideas in *Die räumliche Ordnung der Wirtschaft* (literally the spatial organization of the economy). Due to the language it was written in and the outbreak of the Second World War, it remained a little known work even though it did contain a number of American examples. It was not until 1954 that an English translation was published under the title *The Economics of Location*. Lösch's ideas soon became mainstream. A systematic academic approach to geography (spatial science) was rapidly gaining in popularity at the time, and in economics, the field of regional science was developing. Lösch's location theory is often contrasted with Alfred Weber's. Weber's theory focuses on finding the cheapest location, while Lösch aims to find the location that yields maximum proceeds. Central to Lösch's ideas about economic space are market forces rather than production and consumption, which he grouped together. To him, economic science was about studying interpersonal relations, e.g. between producers and consumers, and geographic space formed an integral part of the economy. As he wrote in the epilogue to his book, 'If everything occurred at the same time there would be no *development*. If everything existed in the same place there could be no *particularity*. Only space makes possible the particular, which then unfolds in time.' According to Lösch, the boundary of the market for a particular product is the point where the consumer can choose between two equally priced products. Producers can potentially enlarge their markets by adjusting their prices. This is made possible by internal benefits of scale, as they can drive prices down. On the flip side, a larger market invites more competition, and any newcomers will try to steal some of the market. But where will they set up shop? Lösch thought it unlikely that new competitors would settle close to the existing provider since then each of them would only earn as much as the demand in half the market would yield. In this scenario, they might both sustain losses. So competitors would spread out in such a way as to create a regular pattern comparable to Christaller's hexagonal pattern. The exact pattern depends on the extent to which they can generate internal benefits of scale (the larger the benefits the fewer the locations). It also depends on transport costs (the higher the costs per product the larger the number of locations). As the outcomes of this exercise are similar to Christaller's, the two theorists are often discussed in tandem. However, in view of his theoretical depth and emphasis on mutual relations, Lösch merits a separate mention.

2.4 Conclusion

Classical and neoclassical economics deliberately reach beyond reality. Their models are normative in that they make statements about what the world should be rather than what it is. Their theories offer positivist explanations based on presuppositions about reality in order to formulate generally applicable laws. Key assumptions are isotropic space, perfect competition, linear transport costs and lastly the existence of 'economic man,' a fully informed entrepreneur whose decisions are entirely rational. When deciding where to establish his enterprise, he judges locations purely in terms of minimizing costs (classical) and maximizing income (neoclassical). Neoclassical theorists also take consumer behaviour into consideration, not just the costs of input factors. Although the name might suggest otherwise, classical and neoclassical location theory is still current. Producers and consumers still try to minimize costs and maximize use and income, and the decision where to manufacture or buy something often remains a matter of price comparison. What is often held against neoclassical theorists is their use of assumptions, which divorce theoretical reality from actual reality, and their lack of focus on the fundamental uncertainties agents have to operate within. In the light of this criticism, it is worth bearing in mind that neoclassicists are not as interested in explaining why a business is located where it is as in seeking to understand an entrepreneur's underlying economic motivation for settling in a particular location. For a neoclassical locational theorist, the geographic distribution of economic activity is always the outcome, never the starting point (Brakman, Garretsen & Van Marrewijk 2001, p. 323). The price for this focus on economic factors is that institutional, political, social or other aspects of geographic interest are not taken into consideration. They will consequently be addressed in Chapters 3, 4 and 5.

3

A behavioural approach to location choice

3.1 Introduction

In Chapter 2 we introduced classical and neoclassical location theories. There has been criticism of their assumption that agents are always fully informed and all their decisions are rational (*homo economicus*). However, in reality it is impossible for people to have all the relevant spatial information. Agents do not make fully rational decisions on where to set up shop and it is not always the place where the difference between costs and benefits is greatest. Considerations that might appear irrational also play a role in the decision-making process. For example social ties might bind an entrepreneur to a certain region, even though somewhere else would be more logical.

In the 1960 and '70s, in response to the calculating picture of human behaviour painted by classicists and neoclassicists, psychologists developed a behavioural model with people acting on the basis of the available information and not necessarily aiming for the most profitable solution. In 1967 this model of human behaviour was interpreted geographically by Allan Pred (see Section 3.3). The central concept in behavioural economic geography is understanding the location choice process of individual businesses. What are their reasons for choosing a particular location? How do they go about obtaining the relevant information and how do they choose a location? Behavioural location theory allows for emotional ties to a town or region that may influence the decision.

Another aspect distinguishing behavioural theory from classical and neoclassical economics is the analytical process. Classical and neoclassical theories are deductive and seek to understand reality using the model as point of departure. The behavioural model is inductive and generalized conclusions are drawn from specific individual observations. Rooted as it is in psychology, the focus of behaviouralism is firmly on the human being. It is not factories that choose locations, it is their directors, managers, shareholders and/or works councils who decide whether and where to move production. Rather than the outcome of a calculation, the choice of location results from a decision-making process in which personal motives can equally play a role.

3.2 The bounded rationality of the decision-maker

One of the main critics of the concept of the homo economicus was Herbert Simon (1916-2001). A versatile social scientist, he was awarded the Nobel Prize for Economics in 1978. In 1960 he introduced the concept of 'bounded rationality' based on his belief that rationality only played a limited role in human behaviour. To make a fully rational decision, a person would have to identify and organize all the possible alternatives, determine the effects of each alternative, and draw a comparison to establish the most efficient approach. According to Simon this is not only impossible, it is unwise because the costs of gathering that much information and conferring with others would outweigh the benefits, not to mention the inordinate amount of precious time involved. People, Simon said, are not *optimizers* but *satisficers* (a combination of satisfy and suffice). If the outcome of a decision is deemed satisfactory, most people are prepared to accept it. Not everyone strives for the best possible outcome of an economic decision. In Simon's view, the homo economicus should be replaced by the *homo psychologicus.*

The homo psychologicus has to live with incomplete information, uncertainty and risk. In light of this, economic theory is faced with a fundamental problem: How can economic progress be made if neither the producers nor the consumers can make optimal decisions and they have no way of knowing what the consequences of their actions will be? Simon's answer was *docility*, 'the tendency to depend on suggestions, recommendations, persuasion and information obtained through social channels as a major basis for choice' (Simon 1993, p. 156). People are open to advice because they know their bounded rationality prevents them from making an optimal choice. Seeking advice or collaborating generates the exchange of information, which according to Simon is the best way to compensate for bounded rationality.

Psychology is now firmly embedded in economics, witness the fact that the 2002 Nobel Prize for Economics was awarded to psychologist Daniel Kahneman (1934). In collaboration with Amos Tversky, he developed *prospect theory*, with bounded rationality as a key concept. People detest losing something they possess and exhibit *loss aversion*. Repeated experiments have shown that in gambling, people are only willing to risk losing 20 dollars if there is a chance of winning at least 40 dollars. Loss aversion explains why the property market can come to a long-term deadlock: ... no one is prepared to take the loss! A good example of this mechanism is 'A Tale of Two Pizzas' by Levin *et al.* (2002). The authors used an experiment to demonstrate how much difficulty people had parting with what they had. The customers at a pizzeria in Rome and one in the American state of Iowa had a choice: a $5.00 bare-bones pizza they could add any of 12 different toppings to at $0.50 extra a topping, or an $11 super-pizza with all the trimmings and an opportunity to remove the ingredients and subtract $0.50 every time they did so. If the two groups made purely rational choices, they would end up paying roughly the same amount. However, this was not the case. Faced with the super-pizza, they found it hard to part with any of the ingredients, and the bare-bones pizzas were sparsely topped with extras. In the same vein, Henriëtte Prast, a proponent of psychological economics in the Netherlands, advises health insurance companies to offer their clients a top-of-the-range policy and allow them to cross out compensation for particular treatments. The insurance companies can achieve a higher income from premiums this way. The same thing happens on the commercial property market. Developers like to create buildings with extra facilities, and although tenants have the option to specify what they don't want to use, they rarely do. Thanks to the additional facilities, the landlord can charge a higher rent.

Other scientists have also questioned the validity of fully rational behaviour. An experiment by biologist Frans de Waal demonstrated a lack of rationality in the behaviour of capuchin monkeys. As a reward for returning a token to him, he gave them a piece of cucumber. This worked well until one of the monkeys was given a grape instead. Monkeys are far more fond of grapes than cucumber, and from that moment on the other monkeys refused to cooperate, obviously thinking you can keep your cucumber. This makes neither economic nor biological sense, since cucumber is better than nothing at all. A similar mechanism appears to be at play with humans. Imagine the following experiment. A subject is given 100 euros to divide with another subject. If he can persuade the other subject to accept the split, he can keep his share. If he wants to keep too much himself, the other subject will be insulted and would rather have no reward at all, and the first subject would not get any money either. This is also irrational, but illustrates the role emotions play in making choices. De Waal argued in favour of a more evolutionary focus in economic analyses to explain for example the economic crisis that began in 2008. We return to this in Chapter 5.

3.3 Allan Pred's behavioural matrix

The psychological approach to the location issue was further developed by American economic geographer Allan Pred in his *behavioural matrix* (Pred 1967). Pred is considered the man behind the behavioural approach to the location issue. Inspired by the ideas of Swedish geographer Wolpert, Pred emphasized the imperfect nature of knowledge and information. It is not feasible to gather all possible knowledge, and an optimal use of information is often difficult as well.

Box 3.1
Allan Richard Pred (1936-2007)

American geographer Allan Pred is one of the founders of behavioural geography. Born in Bronx, New York as the grandson of Jewish immigrants from Poland, he studied geography at Penn State University and got his PhD at the University of Chicago in four years. Chicago was not only renowned for its Economics Department (see Herbert Simon) but also its Sociology Department. It was here that in the 1920s and '30s, Robert Park started the Human Ecology School, where concepts from biological ecology were applied to the urban environment, i.e. the well-known invasion and succession models of population dynamics in urban neighbourhoods. After his PhD, Pred moved to Berkeley, where he was appointed Full Professor in Geography at the age of 34. His early academic career was marked by theories on the historical development of American cities. In Chicago he studied Herbert Simon's behavioural economics. Pred married a Swedish woman and he and his family lived alternately in Sweden and the United States. In Sweden, Pred read about Julian Wolpert's study of behaviour and perception among farmers in central Sweden. Wolpert demonstrated clearly that the farmers made suboptimal choices and their actions were far from rational. Partly as a result of this theoretical cross-pollination, Pred developed his behavioural matrix (Figure 3.1) in 1967. Also known as Pred's Matrix, it is perhaps the clearest example of how behavioural geography works. Allan Pred's scholarly oeuvre comprises 22 books and over 70 articles in journals.

Well aware of human imperfection, entrepreneurs provide knowledge and information in an organized format, e.g. via special departments at their businesses that provide specific, frequently needed information. As the selection of a new location occurs relatively sporadically, they acquire the necessary information for the specific purpose. Larger businesses might bring in location consultants and smaller ones often call in an estate agent specializing in commercial property.

Small businesses, however, tend to remain in the same municipality or neighbourhood, so they already have all the relevant information about alternative locations. It has been suggested that the behavioural approach should thus only be applied to large businesses, where the ultimate choice is often the outcome of a clash of interests between managers. In addition, in large businesses employees have a say via the works council in the decision whether or not to move. Location choices depend on the businesses' capacity to obtain a large quantity of good-quality information and utilize it in a meaningful way. Businesses want location choices to be simplified and reduced to understandable proportions. To this end, various objectives are ranked (incremental rationality), rules of thumb are developed (procedural rationality) and pref-

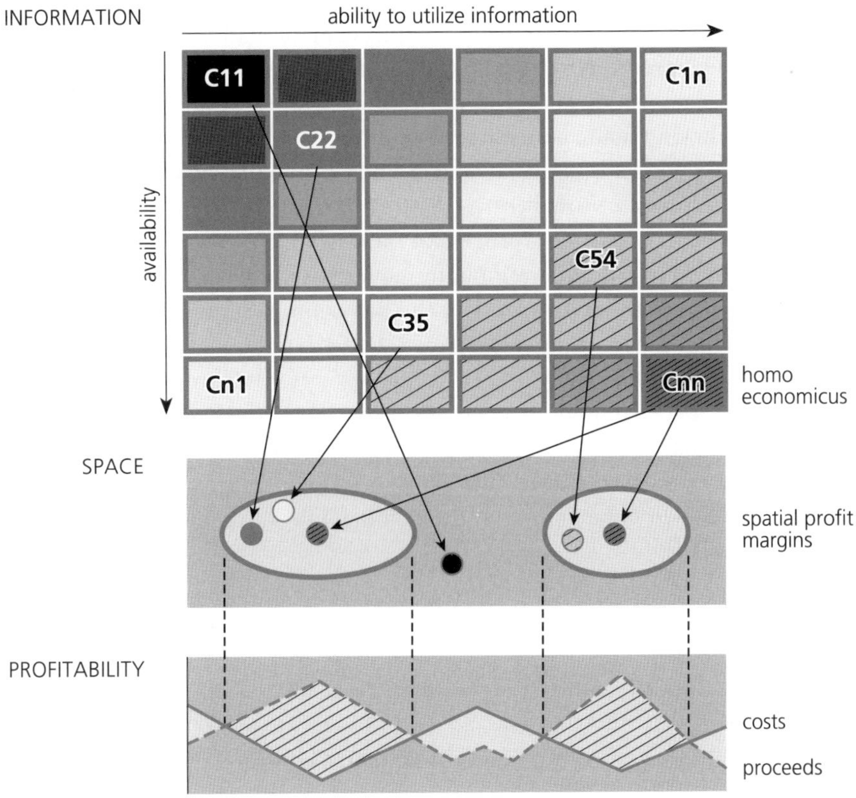

Figure 3.1
Allan Pred's behavioural matrix
Source: Rodrigue 2012

erences are made explicit (expressive rationality). All this implies that the location choice is not only determined by the site qualities, but equally by the personal characteristics and organizational responsibilities of the decision-makers. Different businesses operating at the same location and on the same market can choose different locations when they decide to relocate.

In Pred's behavioural matrix (Figure. 3.1), the views held by businesses change over time as they learn via interaction with the environment, whether the market or other businesses. This enables them to make increasingly balanced decisions. Expressed in terms of the matrix, this means they shift towards the lower right-hand quadrant, where businesses place the neoclassical location theories based on the premise of the omniscient entrepreneur. However, even taking the effects of the learning curve into account, the idea of reaching this quadrant remains purely theoretical.

The main value of the behavioural matrix lies in its usefulness as a conceptual tool. It has rarely been tested empirically. This dearth of robust measurable data makes it hard to place a business in Pred's matrix. Researchers tend to ask entrepreneurs what information and perceptions of the location they based their choice on. Relocating is very costly and has a huge impact on daily routines on the work floor. Businesses do not want to be perceived as having made an irrational location choice. There is often a level of tension after relocation and facts or opinions come to light that are counter to the entrepreneurs' own convictions or opinions, so they rationalize their choices after the event. This process of rationalizing post-event and making the decision fit in with the rational framework is known as 'cognitive dissonance' and is a substantial drawback of the behavioural approach. Its existence has been demonstrated numerous times. For example, a weather forecast predicting sun and showers was seen by two groups. One group went to the beach, the other stayed home. The people who went to the beach said the weather wasn't really that bad (look, the sun is coming out again), and the ones who stayed home focused on the rain (oh no, not *another* shower). Both groups rationalized their behaviour after the event, as is also frequently the case with location decisions.

Box 3.2

Europe's leading cities: London, Paris and Frankfurt

British commercial property consultants Cushman & Wakefield are a good example of the rise of behaviouralism in research. Every year since 1990, they have asked 500 large companies which European cities they thought had the best entrepreneurial climate. As was also the case twenty years ago, London, Paris and Frankfurt come out on top, the cities the companies' managing directors are most familiar with. A total of 82% know their way around London, 74% are at home in Paris, 62% in Frankfurt and a mere 16% is familiar with Athens, 17% with Oslo and 18% with Helsinki. The size of the city also appears to be a factor. The main shifts have taken place among the cities ranked below the top four. The popularity of cities in Germany (Berlin, Munich) and Spain (Barcelona, Madrid) is growing. In particular Berlin and Barcelona have attracted the attention of the managing directors. The same was true of London as it prepared to host the 2012 Olympics. Central European cities like Zurich, Geneva and Milan are now lower on the list.
For more than 50% of the managing directors, four essential factors shape their preferences regarding European cities:

- access to markets (61%)
- qualified staff (58%)
- telecommunications quality (55%)
- fast international connections to other cities (51%)

For all four location factors, the top three cities score highest. Strikingly, the cost of office staff as a factor only scores 33%. Despite a favourable business climate in this respect, cities like Warsaw, Bratislava, Lisbon, Istanbul and Bucharest still have a low ranking. The same applies to the quality-price ratio of office space, where Leeds, Lisbon, Berlin, Glasgow and Birmingham top the list. It would appear that at the top of the international business community, quality takes precedence over costs. To a lesser extent, this also applies to quality of life. From this perspective, Barcelona, Munich, Stockholm, Paris and Zurich are firm favourites. In 2010, Amsterdam climbed two places. According to the researchers, the Dutch capital had a high score in terms of accessibility and quality of life.

The 10 most popular European cities

	2011	2012	2013
London	1	1	1
Paris	2	2	2
Frankfurt	3	3	3
Amsterdam	4	6	5
Berlin	5	7	15
Barcelona	6	5	11
Madrid	7	8	17
Brussels	8	4	4
Munich	9	9	12
Zurich	10	13	6

Source: Cushman & Wakefield 2011

3.4 Industrial relocation

In the Netherlands, the main proponent of behaviourism in economic geography is Professor Piet Pellenbarg. Having observed how dissatisfied many businesses were with where they were based, he investigated the reasons for their dissatisfaction and their decision-making processes with regard to relocation (Pellenbarg 1985, 205).

Box 3.3
Industrial relocation in the Netherlands

In 2007, the Netherlands Institute for Spatial Research published an overview of industrial relocation in the Netherlands. The data below were taken from this study. In the Netherlands, 18,000 businesses and institutions move annually, representing 4% of all the economic activity and 3% of all the jobs. Most do not move very far: 75% remain in the same municipality and 94% in the same province. Of the 25%

that move from one municipality to another, as many go from a central urban environment to a rural one as vice versa. The industrial migration balance between central cities and the surrounding countryside is practically zero, and industrial relocation streams are balanced between regions as well. There is no mass exodus of businesses from the conurbation in the west of the country known as the Randstad. Growth in employment and the likelihood of industrial relocation are closely associated, although the causality of the relationship is not entirely clear. Sometimes growth prompts businesses to move, sometimes businesses grow as a result of moving. The contribution to local employment of existing businesses newly moved into an area matches that of newly launched companies.

Source: Institute for Spatial Research 2007

The first stage of behavioural research into industrial relocation pertains to determining the *place utility* of the current location. This concept, developed by geographer Julian Wolpert, reflects the extent to which businesses are satisfied with their current location, i.e. the extent to which the location enables agents to achieve certain objectives. For residents of a certain location, this is expressed as how enjoyable it is to live there, but for businesses, it is all about economic results. Seen in this light, place utility represents the extent to which a location meets the demands of the businesses based there. Wolpert (1963) emphasizes the relative nature of the concept of place utility; residents or entrepreneurs compare and contrast the place utility of their location with that of alternative locations. Place utility is a function of what other locations have to offer. If the place utility of the current location decreases or if more favourable alternatives are available, a business will give the possibility of relocation more serious thought, increasing the likelihood of a move. There is however another possibility, i.e. that the business might try to increase the current location's place utility. This can be achieved by improving accessibility in collaboration with the local authorities. Businesses usually don't seriously start thinking about moving until it is clear no solutions can be found for the problems posed by the current location. In practice, place utility has a wide margin.

Once a business decides to move, the process of finding a new location begins. Research into this process generally focuses on motivation factors. A distinction can be drawn between *push and pull* factors as well as factors that motivate businesses to stay where they are. Push factors motivate a company to leave an area, e.g. if the current location does not allow for expansion or is no longer sufficiently representative. A new location's pull factors motivate a company to relocate there. Examples would be good transport connections, including public transport, good parking facilities and the availability of subsidies. There are also factors that motivate businesses to stay in their current location, reasons to stay put after all, having considered all the alternatives. These factors focus on the drawbacks to moving. An entrepreneur's fears of losing workers and of uncertain market prospects are powerful factors in this respect.

Relocation motivation factors change over time, although many remain the same. Pellenbarg (2005) conducted three similar studies among fifty recently relocated companies in the Netherlands of varying sizes and from various industrial sectors. He drew various conclusions from the studies. The main push factor remains a lack of expansion opportunities. In terms of pull factors, the size of a location is seen as less important than the quality, particularly in terms of representativeness. This can be interpreted as a reflection of the ongoing growth of the service economy. The importance to entrepreneurs of being easily accessible to their buyers points in the same direction.

The businesses' knowledge and perception of the current and potential new environments play an important role in the decision-making process. This knowledge or *spatial cognition* goes beyond the motivation factors and requirements of the business. Pellenbarg defines it as 'the basis for judging particular spatial units' (1985, p. 179). To him, the gathering of spatial information is an essentially objective example of bounded rationality. In industrial relocation, spatial cognition is based on how potential new sites are judged. This is linked to the *mental map* (perceptions of the living space) and *regional image* concepts.

A *mental map* represents the picture conveyed by a certain place or region. It can be influenced by factors such as sex and educational level. Whether in the service or manufacturing industry, businesses focus on various aspects when forming a mental picture of a locality. Manufacturers may look at the price of semi-finished products and those in the service sector are more interested in customer accessibility. People with similar backgrounds often have very similar mental maps. They may lead to stereotyping in the form of a regional image, which can be either positive or negative. Examples of stereotyping are *the far north* and *inhospitable Siberia*, or to give some Dutch examples, *remote Zeeland* or *the overcrowded west*. Nowadays, authorities proactively try to develop positive regional images as part of their city marketing schemes (see Chapter 7). A well-known example is the slogan coined to boost the image of the northernmost province of the Netherlands traditionally viewed as remote and rather dour: 'Nothing can top Groningen.' Other examples include 'Utrecht, knowledge and culture centre' for the historic university town, and 'Tilburg, a modern industrial town' for the southern Dutch town that once boasted a flourishing textile industry.

A regional image is generally characterized by a subjective interpretation of an objective reality. These interpretations, however, are also used by businesses and are part of the information set available to them at that particular point in time. As Herbert Simon reminded us, people can only base their rational behaviour on the limited, often biased information available to them. So for an entrepreneur planning a move, certain personal characteristics play an important role, like his level of ambition, his culturally determined values and his cognitively determined mental maps and regional images. Using advertising campaigns, the authorities attempt to anticipate and capitalize on them.

Another aspect of spatial cognition is the decision-maker's experience. Particularly in the case of remote regions, perceptions tend to be subjectively biased (prejudices, stereotypes). The average entrepreneur is poorly informed about these regions. He is likely to be much better informed about the region he is currently based in, since that is where his experience comes from. This explains the *neighbourhood effect*, a tendency to value the familiar and nearby more than the far off and unknown. This effect is evident in the Eurovision Song Contest, where judges usually award more points to performers from neighbouring countries or those they have cultural ties with. In the same way, Dutch entrepreneurs value their own and nearby regions more highly than those further away.

Box 3.4

Location preferences of Dutch entrepreneurs in 1983 and 2003

Between 1983 and 2003, Pellenbarg's research team mapped the preferred locations of Dutch businesses in three comparable surveys (Pellenbarg & Meester 2007). On each occasion, a map of the Netherlands containing seventy locations was presented to 1800 randomly selected businesses. The businesses were from the manufacturing, wholesale, building, transport and service industries, had a min-

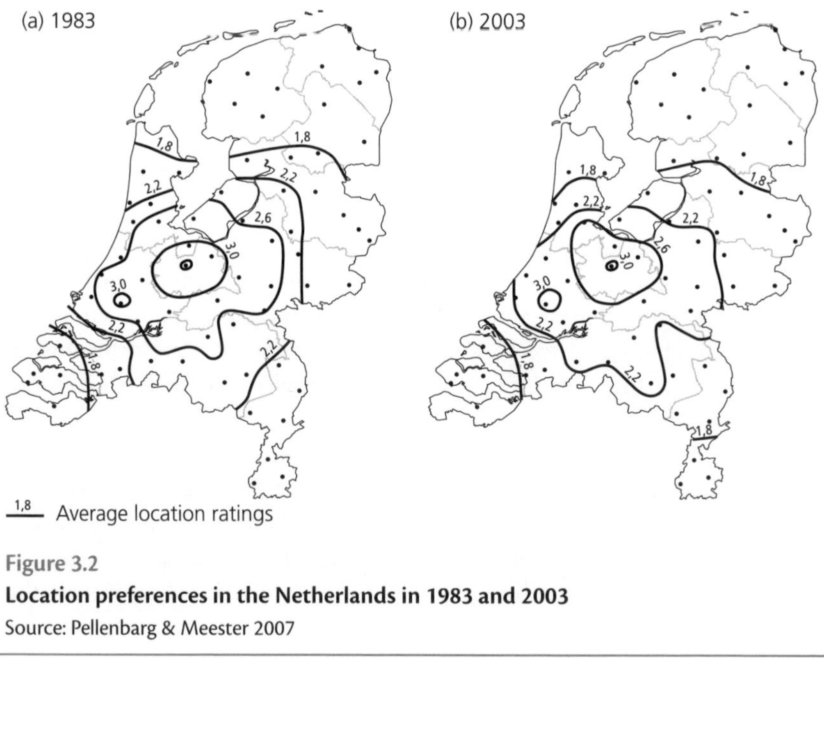

Figure 3.2

Location preferences in the Netherlands in 1983 and 2003

Source: Pellenbarg & Meester 2007

imum staff of ten, and only operated on the Dutch market. The investigators asked how they would rate each of the seventy places if they should need to relocate. A very stable picture emerged. Centrally located Utrecht and the surrounding region were consistently rated the highest, with ratings dropping towards the peripheral regions. These results are in line with behavioural theory. Wherever they are based, businesses are most familiar with the central regions and peripheral areas such as the northern or southern provinces are considered less attractive (neighbourhood effect). In 2012 Pellenbarg repeated the study with more or less the same results as in 1983 and 2003.

Over time it has become increasingly tricky to determine what the dominant location factors are, especially since the 'spatial margins of profitability' are wider than in the past. Developments such as technological innovations, e.g. in transport, give businesses a wider choice of locations they can trade profitably from. As is noted in Chapter 1, many businesses view the surroundings as one large urban field and it makes no difference to them whether they are based in The Hague or the slightly less central Arnhem. In choosing a location for his business, a modern entrepreneur can take more factors into account than purely financial and economic ones without much of a risk. With location factors less spatially top-heavy, entrepreneurs can now also consider other aspects, such as the quality of the living environment, childhood memories of where they were born or used to live, or the location-related behaviour of other entrepreneurs, some of whom they may be friendly with. Some researchers feel it is the soft and not the hard location factors that determine the ultimate location choice.

Box 3.5

Hard and soft location factors

For his Master's thesis at the Amsterdam School of Real Estate, Jeroen Jansen conducted a literature study of location factors. He compiled a list of 91 factors, which he divided into 80 hard and 11 soft factors. Frequently cited hard factors include the size of the premises, accessibility and the proximity of businesses of a similar type. Jansen lists the following soft factors:

Premises:

1 Prestige, representativeness, distinctive architecture

Immediate surroundings:

2 General character and liveliness of the locality

3 Attractively situated

Social enonomy:
4 Work ethic and general mentality of the local population
5 Internationally oriented
6 Reputation of the region
Living environment:
7 Attractive landscape
8 Visually appealing buildings and structures
Local authorities:
9 Accessible, trustworthy and vigorous
10 Quality of the information provided
Other:
11 Personal motivations

Like hard factors, individual soft factors can vary in importance from one firm or industrial sector to another. The research bureau Stec (2001) differentiates between hard factors such as accessibility and labour market and soft factors such as image and personal preferences. According to Stec, the importance of the two types of location factors varies per key activity.

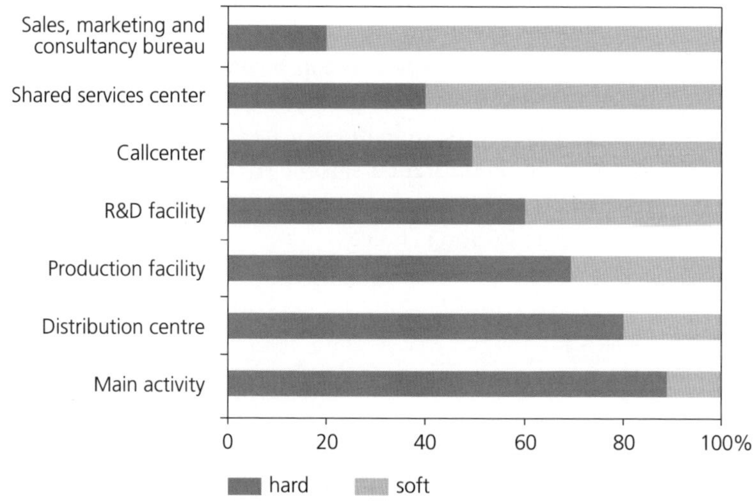

Figure 3.3
Hard and soft location factors per key activity
Source: Stec 2001

Wim Meester (1999) observed that soft factors are often underrated. Survey respondents often give socially desirable answers. Following a choice of location, they tend to state that the more rational, 'hard' location factors were behind their preference. For a researcher to find out whether soft factors genuinely play a role or not, a face-to-face interview is usually required. To make a truly rational choice, an entrepreneur would have to investigate every single place for its suitability as a new location. In reality, however, he does not travel all over the world to compare all the potential new locations. The behavioural approach says he will base his choice on a limited selection of locations and choose the one he feels best meets his entrepreneurial aspirations. Entrepreneurs are only human, and their choices are not solely motivated by profits. Their behaviour is equally driven by their personal and professional circumstances, experiences and social environment.

The behavioural approach focuses on decision-making and the explicit and implicit motivations involved in the process, but it is often criticized for not devoting enough attention to the context decisions are made in. Behavioural economic geography examines how decisions are made from the inside out, and reveals that decisions are often influenced by power relations or government policy. Subsidiaries of large companies, for example, are often tied to the apron strings of the parent company in their decision-making, and farms are sometimes shut down to make way for housing development. In his 2002 thesis, Cees Jan Pen propagated the notion of 'cross-pollination' between behavioural economics and institutional economics, as is discussed in Chapter 4. According to him, industrial relocation can only be understood as part of the overall strategic management of a company. Pen followed Herbert Simon's theories in that bounded rationality can be solved through internal organization. In addition to emphasizing the organizational aspect, Pen also stressed the importance of time in the process of finding the right business location. Location decisions are often made after a protracted process of careful deliberation, during which motivations and considerations can change.

3.5 Conclusion

The behavioural approach to location was a reaction to the normative nature of classical location theories. For many businesses, deciding where to set up shop is largely historically determined, but a quarter of them nonetheless move every year (Pellenbarg 2005). A short-distance move can also constitute a drastic change that businesses do not make overnight decisions about. As it is impossible for them to obtain all the possible information on all the possible locations, they need other ways to find the relevant information. This can be done either internally or externally, e.g. via a consultancy agency. Behavioural economics has demonstrated that the same business environment can be rated differently by different businesses. This is also the weakness of this approach since it yields largely descriptive studies, and more interpretation than explanation. The reasons businesses rate the same locations differently often remain elusive. The consequences of the differences are also hard to explain. Most businesses are not at ideal locations. Though the locations meet some of their requirements, they inevitably have drawbacks as well. For his study on behavioural geography, Pen (2002) conducted an electronic survey among international experts, demonstrating that the number of economic geographic research projects with a behavioural approach had declined. Each scientific approach, the experts said, has its own life cycle. After a while an approach becomes unfashionable and, in accordance with a time-honoured academic tradition, a new theoretical trend takes over. Pellenbarg and Meester (2005) wondered how it is possible that in economics, behaviourism is clearly on the rise, but in economic geography there are so few proponents left. In Chapters 4 and 5 we show how certain aspects of the behavioural approach are still in active use, namely in institutional and evolutionary theories, which are currently growing in popularity in economic geography.

4

An institutional approach to location choice

4.1 Introduction

Agents operate in a context of uncertainty. Businesses and consumers can never be sure of the best price, and the actions of other buyers and sellers can create unexpected situations. Moreover, there is no guarantee a new product will take off. Uncertainties create risks, and to contain them, agents need regulation in the form of norms and customs or legislation. This leads to a new uncertainty: Are the rules properly applied? To limit the risks, agents incur extra transaction costs, e.g. when they draw up elaborate contracts. Rules and regulations and how they are applied vary from one country or region to the next, impacting businesses' location choices. The rules and regulations are called 'institutions' with norms and customs as the informal institutions and laws the formal ones. The main institutions in economics pertain to ownership. They play a vital role in trade, and only after ownership issues are sorted out do products change hands. Institutions are there to prevent opportunistic market behaviour promoting one's own interests at the expense of others. Their aim is to ensure that everyone honours their promises and does not mislead others by altering or withholding information. In short, institutions keep stakeholders straightforward and focused.

In economics, institutions serve two purposes. Firstly, buyers and sellers do not have to make random decisions. Agreements underlie everything on a day-to-day basis and economic activity is largely routine-based. An important aspect is cost-effectiveness. Secondly, institutions serve the communal interests. Society has to be able to rely on rules being followed and laws obeyed. So institutions come with sanctions. They function the same way as traffic rules, without which there would be far more accidents and traffic wouldn't flow as well. We live in a highly regulated society, though the nature and extent of this regulation vary per country and region.

The institutional approach was not developed in economics or geography, but in business management and sociology. In business management, the rules pertain to managerial decision-making in the production of goods or services, and in customer service. In sociology, the focus is on the regulating effect of customs, norms and conventions, and the interaction of agents (buyers and sellers). The difference between the institutional and the classical or neoclassical approach to economics is that the latter is contextual. According to classical or neoclassical theorists, agents only collaborate if it leads to financial gain. In addition, they believe there is no connection between cooperation and the socio-political context it takes place in, since in their view agents are primarily led by market forces. According to proponents of the institutional approach, economic activity cannot be seen as separate from its context. A shift towards sustainability can be the result of corporate values or government incentives through legislation. Regulation determines the context economic agents function in. In this chapter, we examine the roles institutions play when companies select a location. We demonstrate the extent to which rules and regulation play a role within and around companies with regard to location choices. Firstly, however, we ask: What exactly are institutions?

4.2 Institutions, rules of the economic game

It is noted in Chapter 2 that in classical and neoclassical economics, producers and consumers aim to maximize their profits or consumption and minimize their costs and this is the point of departure for their behaviour. Businesses tend to choose the cheapest location that allows them to serve the largest possible market. Chapter 3 explains that according to behavioural economics, this behaviour of buyers and sellers is only partially rational and largely determined by personal motives. This leads them to choose the locations they are most familiar with or that best match their mental image of the ideal location. Most businesses move within their own municipal district or region. Institutional economics has a different take on why businesses choose certain locations. Proponents of this approach view the economy as a game where the players are the producers and consumers and the prize is maximum profit or consumption. The outcome of the game is the result of the players' ability or inability to make the best possible use of the rules of the game (institutions). Institutions contain sanctions as well as rewards, and players weigh them against each other and try to anticipate the behaviour of other players. We call these 'institutional practices.' Rules change and new players join in, making it a dynamic game, all the more since these factors raise the stakes. Economic behaviour, institutional economics says, is always strategic, never impulsive. Applied to location-related behaviour, it states that businesses choose the locations that best fit their corporate strategy. So location is the direct result of strategy.

The simplest model in this game theory is played by two players, both of whom are fully informed. Their chances of winning are equal, so there is no point to making upfront decisions to collaborate with the other player. Take the game of poker, where players always play against and never with each other. The outcome of the game depends on the strategies of each player, not on the information they have. Theoretically, this identical starting position prompts players to opt for the same optimal strategy. But since they know this will not yield more than half the maximum profit, the temptation is to depart from the original strategy, especially if it confuses the opponent, who then also changes track. The winner is the one who chooses the cleverest alternative strategy. In economic terms, this means that by making optimal use of the rules of the game, buyers and sellers can create advantages for themselves and tailor their strategies accordingly. Institutional economists view private property as the best possible way to ensure that all the players stick to the rules. It is in everyone's interest to do so since it is the best way to protect property. Moreover, since they stimulate cooperation, institutions such as right of ownership have a favourable effect on collective interests. Institutionalists emphasize that this is not only in the collective interest, it is in everyone's individual interest. The assumption is that the sum of the actions of individual players does not necessarily result in the optimal collective outcome. Players who only play to win disadvantage each other. A frequently used example is the prisoner dilemma based on a crime with two suspects. There is not enough evidence to sentence either of them, so the prosecution makes a proposal to each of them without the other knowing. If one suspect betrays the other and the other pleads not guilty, the first one goes free and the other is sentenced to ten years in prison. If both plead guilty, they are both sentenced to five years. If neither of them cooperates and both plead not guilty, they are both sentenced to six months for lack of evidence. So for the two suspects collectively, silence is the best tactic. Individually, however, the best strategy is to betray the other one. Denial of guilt is risky. If the other suspect cooperates, you get a ten-year sentence. Another example is the international negotiations to reduce greenhouse emissions. Game theory has the following interpretation. Rather than find the best way to reduce CO_2 emissions, China and the United States choose the cheapest option: Do nothing. If the US were to invest in finding costly solutions, it would harm its competitive position vis a vis China if China made no effort to reduce its emissions. In addition, China would benefit from the US reduction in CO_2 emissions. If China invested but the US did not, the same would happen. The result is that the 2011 climate change negotiations came to a complete deadlock.

Box 4.1

Institutions according to Douglass North

Together with Robert Fogel, the American Douglass North won the 1993 Nobel Prize for Economics. Best known for his book *Institutions and Economic Growth: An Historical Introduction* (1989), North summarized his ideas in an article entitled 'Institutions' (1991). His main research interest was the historical development of institutions and their significance for the economic growth of nations. He drew a distinction between *formal institutions* (legislation and regulation) and *informal institutions* (norms, values and conventions). This distinction coincides with the one between *external institutions* (imposed by authorities) and *internal institutions* (resulting from collective experience). North saw the process of institutionalization as a necessary evil. As countries modernize, the original barter economy is replaced by a market economy. This loosens the ties between producers and consumers. As informal institutions replace formal ones, relations became less personal. North argued that in the West, there is a close connection between the development of institutions of democratic decision-making and those of market regulation, and they have led to the relatively large economic growth there. Herein lies the economic significance of the modernization process. Broadly speaking, North's theory of modernization goes as follows. Initially, the split between producers and consumers caused by the modernization process led to increased uncertainty for both groups of players on the market. It also provided ample opportunity for free-rider behaviour (consuming without paying). These effects prompted an increased awareness of the need for formal institutions, which were best imposed by democratic governments. In modern societies, there was a strong call for efficient government, even if it meant higher taxes. Producers and consumers traded taxation for regulation. Market regulation went hand in hand with a continual increase in labour divisions between people individually and within businesses as a whole, which became more and more specialized. This specialization coincided with the integration of business activity, leading to the development of large-scale companies. To prevent monopolies, institutions such as competition authorities were founded. In the Netherlands, over a million people now work at these regulatory institutions, and companies such as KPN and Microsoft have become aware of the clout that competition authorities have. International trade was also increasingly bound by regulation, sometimes to prevent protectionism, at other times to protect key industrial sectors such as food production from foreign competition. There was also a steady increase in regulation protecting working conditions, e.g. laws forbidding child labour and regulating working hours. The result of all this regulation is that in modernized countries, the interests of authorities and industry came to run parallel. To avoid a complete convergence of their interests, North advocated a renaissance of informal institutions to act as a brake.

Proponents of the institutional approach focus specifically on how information is passed on, taking into account that not everyone has access to the same information or the same amount of information. Akerlof (1970) wrote a seminal article based on the idea that some products live up to the expectations in every respect and others are duds. He uses the perhaps not entirely randomly chosen example of the car industry and distinguishes four categories: new dream cars (cherries), new cars that turn out to be defective (lemons), and used cars of the same two types. If a car is new, neither the buyer nor the seller know whether it is a cherry or a lemon. In the used-car business, of course the seller knows much more about the vehicle than the buyer and the information is asymmetrical. Though the used-car dealer knows more than the buyer, this is not reflected in the price since that would give the game away. As he is particularly keen to sell the lemons, confirming the image of the dodgy used-car dealer, it could spoil the market. This kind of 'cherry picking' often crops up in other branches as well, for example when insurance companies prefer healthy, low-risk clients to higher-risk ones they devise all sorts of barriers for. The same is observed in public transport, where concessions are easier to sell in densely populated than sparsely populated areas. In short, information about what is really happening on the market is unevenly distributed across agents, who make efforts to eliminate this asymmetry. For businesses this means incurring transaction costs, and they prefer locations where reliable information is abundantly available. Institutional economics can explain the rising importance of credit rating agencies such as Standard & Poor's and Moody's. Governments have grasped this and make every effort to provide information, catch the eye of investors and influence the decision-making process (see Chapter 7 on city marketing).

4.3 Transaction costs, a key aspect of institutional economics

Unlike behaviouralism, institutionalism assumes that people act rationally and that institutions aid producers and consumers in their rational behaviour. For businesses, this means institutions help them minimize the costs of market transactions. Business economist Bart Nooteboom (1994) distinguishes three phases in market transactions:

- *Contact.* A transaction means there are two parties, a supplier and a recipient. Naturally, there has to be some sort of existing relation between the two for the transaction to be possible. This means the recipient has to incur costs to find a supplier and the supplier incurs marketing costs.
- *Contract.* Most transactions come with terms of delivery. To avoid buying a lemon or to ensure payment, the parties draw up a contract, thus incurring contract-related costs.

- *Control.* To make sure the transaction runs smoothly, the two parties agree to carefully monitor how well the agreed terms of delivery are met, leading to evaluation costs.

Box 4.2
Ronald Coase

As early as 1937, British-born economist Ronald Coase published an article entitled 'The Nature of the Firm.' It was to become highly influential, but his ideas only really took off in the 1960s following a steep rise in market transactions. In 1991 he was awarded the Nobel Prize for Economics. Coase's null hypothesis is that producers buy all the necessary input and resources on the market and farm out the marketing. This way a business has no need for any industrial organization. Like a sole trader, it can arrange for goods to be delivered on time from the comfort of his own home The difference between its outgoings and incomings is its profit. After a while, however, in Coase's example, the business discovers its transaction costs are rather high. If it is exclusively dependent on suppliers, it runs the risk of being underinformed or poorly informed, and many different suppliers means many different contracts, each with its own additional costs. Repeat transactions can be better dealt with within the business. Coase sees the business as being confronted with two different allocation mechanisms: externally (on the market) with supply and demand and internally (within the business) with planning and transfer prices. The market price is shrouded in uncertainty, especially if the market is assumed to be characterized by imperfect competition, which is the assumption Coase makes. He believes that no situation of equilibrium on the market will last long since newcomers are constantly entering the market, particularly if profit margins are good and new technologies emerge. Markets are dynamic places. Businesses need to be keenly aware of this and anticipate change in their operational strategies, and the three types of transaction costs flow from this.
Together with transformation and transport costs, transaction costs are an integral part of the production process in a free-market economy. Without transaction costs there would be no commerce, and no need for institutions monitoring compliance with internal and external agreements.

Economist Oliver Williamson (1975) built on the ideas of Coase, drawing particular attention to opportunistic behaviour on the market – the issue of whether partners all invest proportionally in the communal network. The partner who makes the lowest transaction-specific investments can put pressure on the other partners by threatening to leave the collaborative network. For example, if companies decide to jointly develop new aircraft, it is easier for a tyre specialist to leave the partnership than a company specializing in wings, since wing production is limited to aircraft manufacturing. If a business invests in its own machinery to be able to meet its partners' demands, this is a clear sign it is committed to the partnership. This commitment can go too far and the effects of making transaction-cost-specific investments cannot be undone without incurring losses. This poses a problem if a business, prompted by market developments, set its sights on other technologies or products. The partners' rights can be protected via internal or external institutions, or in the extreme scenario by subsuming the partners into a larger organization. Combining property rights in a larger or even a new organization is a tried and tested means of averting risks due to transaction-specific investments. In the event of uncertainty, it allows the business to consider replacing a market transaction by an organizational transaction via the internal organization rather than the external market. An entrepreneur might wonder: Should I buy the required product from a third party or manufacture it myself? (The buy-or-make decision, see Figure 4.1.)

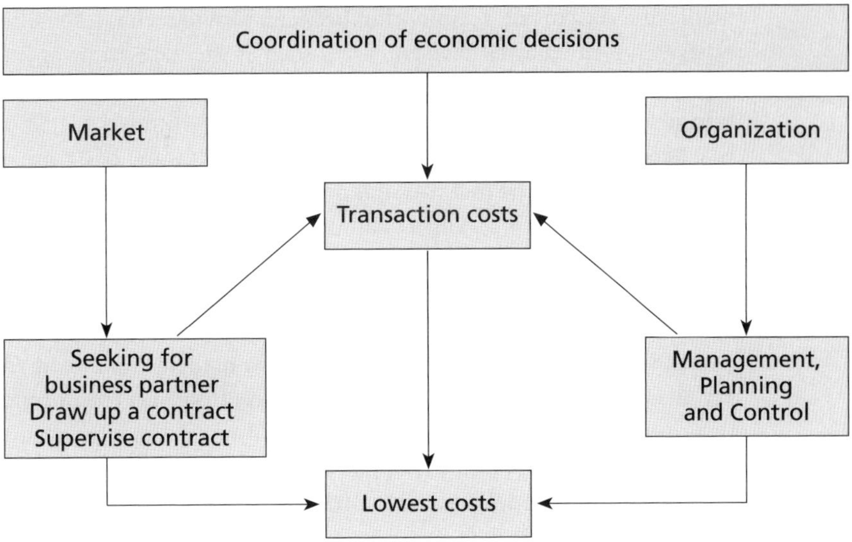

Figure 4.1
A look at transactions costs

Market and investment-related risks may be greatest in volatile or knowledge-intensive markets, and companies need to be able to provide fast and flexible responses to whatever opportunities there are. They are constantly faced with the need to make buy-or-make decisions.

There is a constant question: Do businesses use institutions to guarantee quality or do they cut them out of the process to be more cost effective? Businesses tend to focus more on reducing transaction costs if the number of transactions increases, if there is uncertainty about the direction the market will head, or if there are transaction-specific investments. These are investments made to ensure the smooth operation of a transaction, e.g. extensive screening of potential suppliers or in the case of knowledge-intensive services. Generally speaking, a reduction in transaction costs is considered a positive thing leading to reduced costs and greater efficiency. This said, many businesses survive by providing intermediary services. Eric van Damme (2003) is far from optimistic about the future of the traditional business broker, citing the example of a TV advertisement for Dell computers using the slogan 'We have knocked out the third man,' a reference to Dell's successful practice of not using stores and delivering directly to the customer. Lately however Dell has rethought its approach since not having a presence in stores appeared to sometimes be a disadvantage. Another example is that of the estate agent who sees his businesses dry up because the Internet now enables buyers and sellers to find each other without his intervention. The rise of the Internet substantially reduced consumers' costs of researching their purchases, diminished the importance of distance and location, and put immense pressure on the intermediaries. As a result, competition between providers has intensified and new types of intermediaries have emerged, such as the 'DIY estate agent.'

In 2003, Van Dalen and Van Vuuren wondered what the actual size of the transaction sector was. Oddly enough, only very limited empirical research has been conducted into this sector and transaction costs are not clearly defined. Based on the institutional definition 'costs ensuing from the transfer of rights of ownership,' Van Dalen and Van Vuuren reached the substantial figure of 1.8 million people working in the transaction sector, i.e. nearly 29% of all the people working in the Netherlands. In particular businesses with an international outlook have an ample staff to monitor transactions, which makes sense since long-distance trading can involve significant risks. Transaction costs impact businesses' relations and physical surroundings. Transaction-specific investments help generate lasting relations in outsourcing and knowledge networks. Once they exist, businesses tend to settle in the regions where their main suppliers and knowledge networks are based. There is plenty of space for opportunism in such a turbulent commercial environment. In a highly dynamic market, there is little point to making elaborate contractual arrangements. In addition, the first transaction provides the best chance to act opportunistically. After a while, a collaborative stance is more fruitful than an opportunistic one. When times are turbulent or regular relations are scarce, businesses try to minimize external uncertainty by maximizing internal planning. This makes them more stand-alone, more independent, and shields them from external threats. Conversely, if transactions are not shrouded in uncertainty, firms tend to enter into relations with external partners, making the most of the market mechanism. Allen Scott used Coase and Williamson's transaction cost theory to explain what he calls 'new industrial spaces' (Scott 1988). According to Scott, the need to minimize transaction costs prompts businesses to base themselves in each other's proximity. There are a number of advantages, since this reduces the costs of finding relevant information and allows them to find fast solutions to technological problems, exchange specialist skills, organize their activities in tune with each other's production schedules and so forth. Yet Manshanden (1996) emphasized that the aim of transaction cost theory is not to explain why businesses are located where they are, but first and foremost to gain insight into industrial organization issues. He conceded however that proximity leads to transaction cost reduction, stating that businesses 'swap transport costs for transaction costs' (Manshanden 1996, p. 122). This mainly applies to low-value services. Distance is less of an issue for high-value knowledge-based services where investments are often transaction-specific. Below we look at the practical implications of these theoretical ideas for location choice.

4.4 Institutions and location choice

A well-known example of institutions playing a role in location choice can be found in the work of organizational psychologist Geert Hofstede (1980, 1991), who compared various dimensions of the value systems of 117,000 IBM workers in different countries. Hofstede viewed cultures as the 'collective programming of the mind,' with clear differences in terms of social distance, masculinity, uncertainty avoidance, individualism and long-term vs short-term orientation. In his view, it is advisable for businesses with an international outlook to take these cultural differences into account, and the same holds true for national authorities and the firms they set their sights on. Cultural distinctions also impact the extent to which other cultures value institutions. According to Fukuyama (2011), the 'rule of law' concept, the extent to which societies set store by their legislation and institutions, is of crucial significance. There is a logical link between a robust legal system and low levels of corruption. Causality is also a matter of debate. Do well-developed institutions create growth or does increased wealth lead to better developed institutions?

The perspective of national and regional cultural similarities rather than differences has generated the concept of 'embeddedness,' referring to the connection between close sociocultural ties and flourishing trade relations. Being on friendly terms with another nation makes lengthy negotiations unnecessary, and the partners are better informed on the market developments. This also applies in situations where production is partly outsourced, since known suppliers are treated with greater understanding. In turn, embedded suppliers deliver on time and in accordance with the agreed price and quality without the outsourcing company having to check or monitor them. Sometimes a main contractor and subcontractor join to engage in innovative activity, which

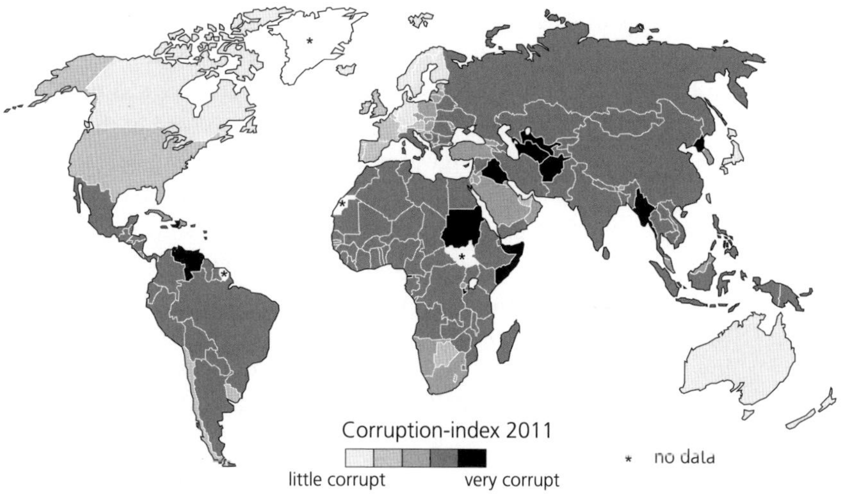

Figure 4.2
The corruption index, an example of institutional differences between countries
Source: *Transparency International 2012*

strengthens the ties between them that they both value so highly. In economic geographic terms, the key aspect of embeddedness is not so much the advantages of reduced operating costs or increased information density, but the greater ease of joint learning experiences and the mutual loyalty it fosters. Learning experiences like this are less frequent in purely commercial relations like those between main contractors and jobbers – the suppliers of simple, standardized products or services. The embeddedness approach may be based on partnerships rather than business contacts and their ensuing costs, but it recognizes the difference between costs and benefits. Firstly, trust-based relations inhibit opportunistic behaviour, thus reducing the need to spend money on transaction costs. Secondly, social ties are a necessity for many small businesses that lack the financial resources to innovate and export. Thirdly, a network based on social relations is generally highly flexible. Since there are relatively few contractual obligations, the business partners can easily adapt the objectives they have agreed upon. Fourthly, social network partners often adopt an open attitude towards each other that fosters collective learning processes. The economic geographic drawback to this social network-based approach to embeddedness is that the local or regional environment is less and less important to their networks. In the business community, the networks encompass the entire globe, particularly in the field of specialist knowledge and skills. In the behavioural approach to the location issue, the emphasis is on why businesses decide to set up shop in a particular place, but institutional economics focuses on how businesses are organized, whether they collaborate with other businesses and with the authorities, and the extent to which the institutional structure of locations matches the investment strategy of the business community. Locations that provide a good match are considered the most highly favoured by businesses. As is noted above, location follows strategy. Most businesses have a strategy specifying what products they intend to bring to which markets, what gives them an edge over their competitors and how they aim to reach their targets. Their strengths or competencies are the point of departure. Whether businesses commit to foreign direct investments depends on their strategy. First they examine their specific competencies. The competitive advantages are at the heart of John Dunning's 'eclectic paradigm' or OLI model (Dunning 1988), according to which three aspects determine whether or not a company will engage in cross-border activity:

- *Ownership-specific advantages (O)*: company-specific know-how determining whether it can hold its own in the face of competition;
- *Location-specific factors (L)*: advantages of the business environment of a country or region;
- *Internalization of ownership-specific advantages (I)*: desire to keep company-specific know-how within the firm rather than relying on alternatives such as licensing.

Essentially the OLI model means foreign direct investment, whether by takeovers or greenfield plants, is only attractive if a company has a viable product to offer, if production conditions are favourable and/or if a company's interests are best served by foreign production. Dunning's point is that there are more factors involved in the decision to invest in a certain country than its business environment or investment climate.

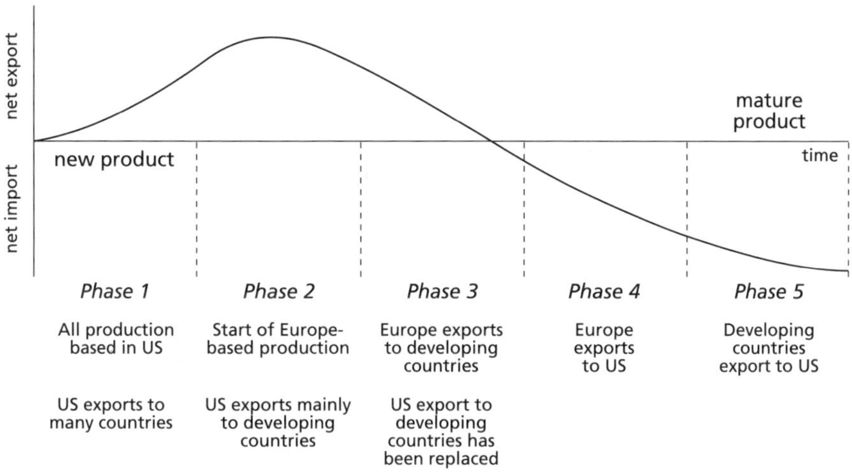

Figure 4.3
FDI and the product life cycle

They are considered the deciding factors in classical and neoclassical economics, but Dunning also includes internal factors (O and I) in his paradigm. It should be noted, however, that once a decision to invest in a country has been made, the location choice will be based on external factors. This leads us to the issue of where to invest, a question also posed by Håkanson (1979) and Vernon (1966). In their view, the answer is simple: Wherever the market is (Håkanson) or production costs are low (Vernon). If a Scandinavian company like IKEA sees investment opportunities on the German market, it will only look for locations in Germany, but to gain access to the EU market, South Korean consumer electronics firm Samsung has a plethora of nations and regions to choose from. This institutional approach to location choice means the question of where to settle cannot be addressed in isolation. It is an integral part of a company's overall business strategy. Following this line of thinking, the chosen location will be the most highly organized one where institutions provide the best business conditions. In choosing a location, businesses consider regional variations in institutions, such as benefits provided by national and regional authorities (tax breaks, installation aid, innovation subsidies, protection against corruption etc.) and a favourable business climate where authorities and other institutions are sensitive to the interests of business and industry.

4.4.1 Institutions and foreign direct investment (FDI)

The size of the investment reflecting a business' level of involvement in a foreign market can vary substantially. Export via direct mailing requires the least investment and full in-house production requires the most. A company envisioning the latter can set up its own new facility (greenfield investment) or take over an existing foreign firm (merger & acquisition), or the parent company can invest in existing facilities abroad (brownfield investment). Companies are apt to commit to FDI (through greenfield investment and acquisitions) if the market there offers attractions and is cost-effective in keeping the cross-border activity within the company. This applies much less to the intervening options in Figure 4.4.

FDI is highly concentrated. Research by British international business and strategy expert Alan Rugman (2005) shows that the 500 largest multinationals account for 90% of all foreign investments, while according to Dutch international economists Brakman, Garretsen and Van Marrewijk (2006), mergers and acquisitions have come to account for nearly 80% of all FDI. Incidentally, they are all takeovers in developed countries. In rapidly developing countries such as Brazil, Russia, India and China (the BRIC countries or Big Four), most investments are still of the greenfield variety. The substantial increase in the number of international mergers and acquisitions has to do with increased

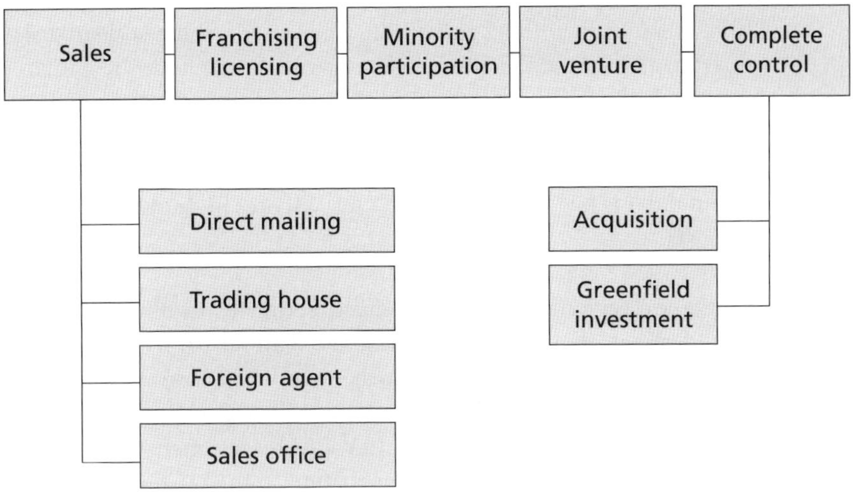

Figure 4.4
Different modes of foreign expansion

liberalization and in particular with the financial markets opening up. Technological innovation also generates acquisitions since young innovative businesses are snapped up by established ones with a strong position on the existing markets. By taking over new businesses, they also acquire the innovations. In addition, takeover trends follow economic trends. In periods of stagnating growth, foreign investments are the first to be slashed, though in growth periods businesses keep an eye on foreign markets. Research by Wall, Burger and Van der Knaap (2010) has also has shown that the global geographic range of FDI expands in periods of economic growth. Via mergers and acquisitions, businesses try to maintain control over the entire production chain (vertical integration), and use them to construct barriers to protect their markets from newcomers. Yet here again, institutions have a role to play. Dutch and European competition authorities guard against excessive integration, leading to monopolistic or oligopolistic markets that could lead in turn to unfair competition. Patent offices are there to protect the intellectual property rights for innovations, giving young innovative businesses a stronger negotiating position should larger ones make a takeover bid. Although businesses mainly look at who they are dealing with, they also note where the potential partner is located. The underlying reasons are generally market or product-related. They could also be motivated by a desire to make use of existing knowledge, gain access to local resources or imitate the behaviour of the main competitors with regard to location. Although in principle the foreign market is a global one, in practice international mergers and takeovers mostly occur between neighbouring countries. Of the approximately 2000 international mergers and takeovers of Dutch businesses in 2010, 55 were with companies in Germany and 53 in the UK. The US was in third place with 52 and Belgium and France came fourth and fifth with 42 and 30. In other words, there is clear 'distance decay' on the international merger scene, suggesting that in this respect as well, institutions have an important role to play, as was also pointed out by Hofstede. Despite a wealth of empirical research, it is difficult to pinpoint which institutions are the most influential when it comes to attracting foreign direct investment. So it is hard to say which country scores highest for which institutional aspects.

There are regional differences within countries in terms of the provenance and destination of foreign investments. Wall (2010) has shown that the dominant regions in Europe are the urbanized ones. Regions dominated by cities like London, Paris, Madrid, Barcelona, Milan, Rome, Munich, Frankfurt and Düsseldorf are the peaks in the European investment landscape. Despite its comparatively small size, the conurbation in the western Netherlands known as the Randstad is a significant player. This is a good illustration of the institutionally characteristic openness of the Dutch economy. A slight qualification is that more investments flow into the North Wing of the Randstad (Amsterdam-Utrecht region) than the South Wing (Rotterdam-The Hague region). The North Wing attracts investments from the US and in Europe, from the Paris region in particular. In terms of inward and outward foreign investments at the European level, Western European urban regions are far more dominant for outward than inward investments. And they maintain their dominant position, with most of the relative changes among regions jostling for positions at the bottom of the list.

4.4.2 Institutions and industrial relocation

According to institutional economics, businesses choose the locations with the most favourable conditions in organizational terms. But what does this mean? A look at relocation and the importance of the institutional aspects makes it clear. In institutional economics, relocation revolves around three questions:

- What strategic regulation is helpful in choosing a new location?
- To what extent does regulation affect the decision-making with regard to relocation?
- Is the anticipated embeddedness in the region an important motivating factor in a business' decision to set up shop somewhere?

Cees Jan Pen (2002) applied the institutional approach to industrial relocation. Like McNee (1960) and Krumme (1969), who laid the foundation for the 'geography of enterprise,' he examined the significance of strategic considerations in relocations and expansions and found that internal considerations are the main driving forces. For example, whether a business expands is linked to the strategic issue of whether its aim is internal growth. If so, sooner or later it may find itself bursting at the seams at its present location. In the Netherlands the state keeps tight control over the industrial location opportunities, largely determining the options open to businesses and thus their relocation behaviour. In addition to the construction of new industrial parks, another important state-related factor in industrial relocations is regulation, e.g. provisos in planning documents. According to Pen, the frequently heard 'lack of space' argument should be seen as directly resulting from businesses' growth strategies, and governments are well advised to look at their actual business strategies. He also notes that decision-making on relocation is more protracted (on average 27 months) than on expansion (17 months). This is mainly because relocations are mostly carried out by larger, growing businesses whose decision-making tends to be somewhat sluggish. In Pen's research, the anticipated institutional embeddedness in the region is addressed via a detailed analysis of two high-profile relocations: the 2002 move of Philips' Domestic Appliances and Personal Care Division from the northern Dutch town of Groningen to the more central Amersfoort on the edge of the Randstad and the move of Grolsch Brewery from Groenlo to Enschede, both in the eastern Netherlands. For Philips, the main reason for the move was to attract and keep highly educated personnel. The Randstad was deemed to have more to offer in that respect. Resistance to the move came mostly from less skilled workers in Groningen, who were of far less strategic importance to Philips. In 2007, the divisional headquarters moved to Amsterdam and in 2008 it was combined with the Consumer Electronics Division to form Consumer Lifestyle. As a business location, Amsterdam fits in with this division's international outlook.

4.5 Conclusion

Institutions play an ever more important role in modern life. With products becoming increasingly complex, there is often an imbalance in the provision of information. It is up to institutions to provide sufficient information. This obviously involves transaction costs and businesses are constantly caught between the need for reliable information and the price of this information. The rise of the Internet and the virtually limitless opportunities it offers for sourcing information have sharpened the lines of this conflict. What it means in practice is that businesses like to stay put, especially if they have good relations with local and regional policy-makers. If they do move, they look for locations with the best institutional match for their business needs.

5

An evolutionary approach to location choice

5.1 Introduction

Charles Darwin's theory of evolution is widely known, yet it may come as a surprise that the evolutionary approach is increasingly popular in economics. Darwin's key concepts are adaptation and inheritance and in economics, Darwin's notion of 'survival of the fittest' means businesses that adapt most successfully to changing circumstances have the best chance of survival. This ability to adapt is passed on to other businesses, leading to changes in the numbers and types of 'populations.' As in nature, there is a great deal of variation in the economy with numerous different survival strategies and starting points. Selection occurs via markets and institutions. Businesses that respond most effectively to changes in competitive circumstances and regulation are most likely to be economically successful. Innovation (commercial renewal) and imitation (learning from each other) play key roles in evolutionary economics. At the population level, they trigger the emergence of new types of businesses and business activities. Forty years ago, there were no ICT businesses in the Netherlands and numerous shoe factories. Incidentally, the emergence of a new sector does not mean every business in the sector will be successful. On the contrary, businesses in new sectors tend to engage in fierce competition with each other. Sooner or later, every new sector experiences a 'shakeout' (see Figure 5.1). At the beginning of the 21st century, many ICT businesses went under. Innovation remains a matter of trial and error and businesses have no way of knowing whether their new product will succeed.

In business, the concept of inheritance means successful adaptations are formalized by being incorporated into how the business is run. We encounter these routines in the previous chapter. Institutions provide a way of dealing with uncertainty. Businesses often incorporate them into their internal procedures, and their fear of the competition makes them guard them jealously. Routines can be seen as a business' DNA and are often described as 'competencies' (see also Chapter 4). Competencies comprise the knowledge and skills that accumulate in businesses over time, and they are the weapons businesses

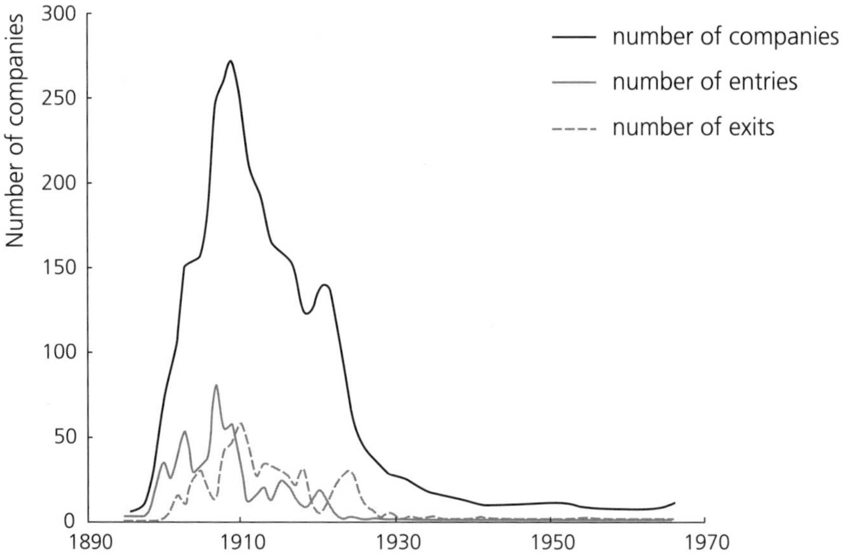

Figure 5.1
The shakeout phenomenon in the history of the US car industry
Source: Klepper & Simons 1997, p. 388

use in the economic selection process. The routines are passed on via imitation and inheritance. Examples of inheritance are spin-offs, I.e. subsidiaries created with the help of the parent company. Spin-offs often focus on a particular competency that has insufficient growth space in the parent company's routines. A highly successful example of this is microchip manufacturer ASML, a Philips spin-off.

Evolutionary economics not only explains location choices from the perspective of adaptation and inheritance, it also refers to coincidence. As in behavioural economics, agents are seen as only partially rational. As is noted in Chapter 3, according to proponents of this approach, businesses are content with options that classicists or neoclassicists would view as suboptimal. Due to the tendency to aim for the satisfactory rather than the optimal, the spatial margins for a location are fairly wide, and businesses would even be able to survive beyond them. Evolutionary economics builds on this insight and gives it a dynamic interpretation.

5.2 New business activity: Chance and agglomeration

In economics, renewal emerges as innovation (see Box 5.1). It can be innovation within existing businesses or the creation of new ones and economic activity. This entrepreneurial view of innovation was introduced by well-known economist Joseph Schumpeter.

Box 5.1
Joseph Schumpeter

Figure 5.2
Joseph Schumpeter (1883-1950)
Source: Karel Franzens Universität Graz

Economist Joseph Alois Schumpeter from the Czech part of the Austro-Hungarian Empire is generally viewed as having been instrumental in the development of the evolutionary approach to economics. His concept of 'creative destruction' is widely regarded as one of its key aspects. Schumpeter compared the economy to a forest where storms blow down the weak trees, making way for new saplings. In due course, they too grow old and weak and get blown over. This is the cycle of 'the perennial gale of creative destruction.'

Creative destruction occurs if there is an important new scientific discovery, a crisis, sudden competition or innovation. For a long time, Canon, Nikon, Fuji and Kodak were each other's main rivals. In recent years, cameras in smartphones have turned Siemens and Apple into formidable competitors as well.

The shifts from agrarian to industrial to service and knowledge-oriented economies have transformed countries, and Schumpeter was the first to view innovation and entrepreneurship as sources of progress. A product of his time, he saw them both as essential ingredients, even the driving forces of a capitalist economy. Through innovation, entrepreneurs create a continual process of creative destruction that helps strengthen the economy.

Along with Karl Marx, Schumpeter was the first economist to devote attention to the dynamics of the free market, noting that the capitalist system is never in a state

> of equilibrium. As his work appeared in the early 20th century at a time of severe economic crisis, his argument that crises are necessary to reinvigorate economies was not always enthusiastically received.
> In 1918 he was appointed Finance Minister of Austria. Austria was in dire need of reorganization and Schumpeter proposed a range of options, raising taxes, forcing banks to buy government bonds, selling the Austrian gold reserves or making drastic budget cuts. Austria had been forced to pay massive war reparations and Schumpeter felt its only option was to print more money, as is now called 'quantitative easing.' The resulting rampant inflation made him highly unpopular and he was forced to resign.
> It was not until the new millennium that his ideas began to enter mainstream thinking. The rapid growth of businesses like Facebook, Amazon and Google had a drastic effect on the world economy, and Schumpeter's prediction that capitalism would ultimately be the victim of its own success is especially relevant at a time when national governments are having to prop up much of the banking sector. He prophesized that if something like this occurred, people would turn their backs on free-market economics and call for more state involvement.
> Source: Driessche 2010; Jansen 2011

In the evolutionary approach, when and where innovative activity occurs is not a given. It is often a matter of pure chance and has been referred to as 'windows of locational opportunity' (Storper & Walker 1989, Boschma 1997). Accordingly, the development of innovative activity in a certain location unfolds in two key stages, an initial phase when the innovations emerge and a second phase, building on the initial phase, when they establish themselves and prove their worth. In the first phase, chance is a factor since it is uncertain which businesses will innovate and which will not. Chances are they will be spin-offs of entrepreneurs who have earned their stripes. Evolutionary economics holds that innovative activity builds on existing activity and does not appear out of nowhere. The likelihood of innovation occurring anywhere depends on the regional distribution of existing businesses and industrial sectors, but it is impossible to predict where it will happen and how many businesses will be involved. If there are no particular location-specific requirements, it can occur at any number of locations.

Evolutionary economics assumes that in the second phase, innovative entrepreneurs can only rely on existing conditions to a limited extent and have to create new ones (new skills, knowledge, markets etc.). They cannot simply use the existing local or regional business environment and need to deploy their own creative capabilities.

There can only be a clustering of innovative activities if innovative businesses have developed into a fully fledged sector (see Chapter 8), making it hard to predict whether a particular business or innovation will be successful. However, the subsequent processes can be predicted as they are 'path dependent.' For example, chance played an important role in Philips' success as a manufacturer of light bulbs and televisions, but it is no coincidence that the Dutch electronics industry clustered around Philips in Eindhoven. The same applies to Steve Jobs' invention of the personal computer in his parents' garage and the clustering of high-tech firms in Silicon Valley, the region where his parents lived (see Figure 5.3).

Clusters develop if there are local benefits for businesses competing with and stimulating each other. The name and reputation of the locality can also be a benefit. Classical economists would say clustering yields tangible results for businesses, driving costs down and profits up. An evolutionary economist would be more interested in the dynamics of knowledge exchange among businesses and the accompanying learning curves. Businesses that can learn from each other seek each other's company by basing themselves in the same vicinity (clustering) or collaborating from a distance (networking).

Figure 5.3
Garage in Los Altos where Steve Jobs and Stephen Wozniak built the Apple 1, the first personal computer, in 1976
Source: Cicorp.com

Richard Nelson and Sidney Winter linked the role and behaviour of small, innovative businesses to the sluggish decision-making of large multinationals. In *An Evolutionary Theory of Economic Change,* their well-known book published in 1982, they cited the ideas of Herbert Simon, the founder of the behavioural approach to location theory (see Chapter 3), and of Schumpeter. In a nutshell, Nelson and Winter's theory is as follows:

- In businesses, decisions are always made in the context of a certain starting situation. As this situation influences the decision-makers' perception, it sets the boundaries their bounded rationality takes shape in.
- Most decisions made within a business are routine ones governed by routine behaviour. Like all humans, entrepreneurs avoid taking risks, prefer-

ring to walk the well-trodden paths of well-defined decision-making procedures. Changes in this routine behaviour, whether enforced or not, are often perceived as threatening.
- Despite the dominant role of routine, businesses have to be on guard. They may not be open to change, but the world around them changes constantly. To hold their own in the face of competition, they will have to innovate. Even if they do engage in change by innovating, most of the accompanying decisions are still routine ones.

As a result businesses, especially large ones with a slow pace of change, gradually adapt to changing circumstances and evolve. This is why dynamic young people who develop something new at large corporations are apt to break away and set up their own businesses, the spin-offs alluded to above. Large corporations enable them to learn the ropes and scout out the market, but do not give them enough space to fully develop their creativity. They spot opportunities that elude their employers and this scenario can prompt an 'innovative spin-off.' Or a young entrepreneur might feel the existing market is large enough to accommodate more businesses of the same type and start a smaller version of the parent enterprise, stealing part of its market. We could call this an 'imitative spin-off,' but in practice it is hard to distinguish between the two types.

Box 5.2
Spin-offs and start-ups: different geography

In his 2006 thesis, Sjoerd Jan Koster demonstrated how the regional distribution of spin-offs in the Netherlands differs from that of regular start-ups (see Figure 5.4). There are numerous spin-offs in the southern province of North Brabant and the Amsterdam and The Hague regions, though start-ups appear to prefer more central locations such as the Utrecht region. In addition, there is much more regional location variation in spin-offs than start-ups, suggesting that regional conditions are more important for spin-offs. One such condition might be a large concentration of successful innovative businesses that can serve as parent companies. The presence of many similar small businesses (earlier spin-offs) can also be a factor.

In the context of location choice, the most relevant evolutionary aspect is that there is ample stability in the regional pattern of economic growth and activity. Once a region has become a core region, it tends to maintain this position. There has however been a paradigm shift from primarily geographical to more relational in economic geography as a result of the recent rise in networking (Bathelt & Glücker 2001). Michael Storper (1997) viewed modern economic geography as the study of interconnections between three key concepts, tech-

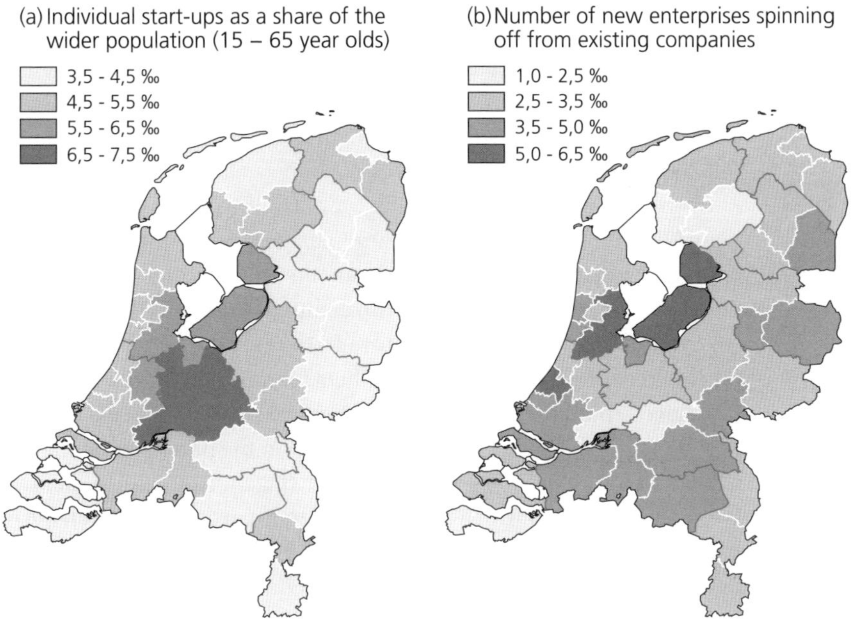

Figure 5.4
The regional distribution of start-ups and spin-offs in the Netherlands
Source: Koster 2006

nology, organization and territory. The interconnections have less to do with costs or financial economic relations than with 'untraded interdependencies' i.e. non-tradable, often implicit knowledge or skills that do not easily transfer, as in informal institutions and social or cultural conventions. Randstad, an employment agency for temps, encountered this phenomenon when it set up branches in France. The offices were furnished and the French labour market looked promising but nothing happened. The Dutch management assumed French employees would show initiative and simply get on with it, but instead they were waiting for instructions from their superiors.

When it comes to knowledge or skills that can only be transferred face-to-face, pleasant locations where people can meet comfortably are by far the preferred venues, and many cities invest in stylish premises. A good example is the Amsterdam business district known as the South Axis or the Financial Mile. As a consultant based there puts it, 'There is a real business vibe here.'

As is noted above, networking is not new in economic geography, we have long known that 'no business is an island.' Smaller businesses have to call on each other more often than large multinationals, the latter being in a better position to internalize numerous relations (see Dunning's OLI model in Section 4.3). However, many large businesses are returning to their core activities as part of their strategies. To them, maintaining relations with other businesses is vital. For practical reasons, businesses often network with regional counterparts (regional networks). If the supply of a vital part, e.g. a car engine, stagnates, it can jeopardize the entire production line and thus the entire enterprise, leaving the director with no other option than to send his workers home temporarily. Working with suppliers in the same region reduces the danger of this happening.

Regional networks are not only important to small enterprises. René Wintjens (2001) demonstrated how large Japanese and American businesses gradually developed regional networks in the Netherlands. In particular, foreign businesses based in the south of the country obtain goods, services and knowledge right there in the region. Naturally, the extent to which this occurs depends on the freedom they are given by their foreign parent businesses. This can be explained from an evolutionary perspective (Boschma *et al.* 2002). Opening a new branch means uncertainty, so it relies on the parent company and procures its staff and supplies internally. At a later stage it sources products and services locally. The extent to which it does so depends partly on how well the new branch is doing. Profitable branches are often given more freedom to develop regional networks and generally do not hesitate to do so. The ones that do exceptionally well continue to grow, sometimes even outgrowing the region. Successful businesses can become integrated in Europe-wide networks and continue to tap into their own regional labour and knowledge markets, playing a regional as well as an international business board game, as it were.

The Kondratiev cycle with its long economic waves (see Chapter 9) is not so much about cyclical as about structural change. The cycle of increasing and decreasing productivity (business cycles) begins with key innovations such as the steam engine that triggered the Industrial Revolution. This is an extreme form of evolution, since the existing economic equilibrium was disturbed by a minority of innovative businesses (early adopters), who introduced new ways of working. Innovations like the steam engine are called 'radical innovations' and lead to entirely new production regimes. By contrast, 'incremental innovations' are relatively small modifications to existing technology though their impact can be great. Since many other businesses soon imitate the innovations, they can have a cumulative effect. Research has been conducted by evolutionary economists on the diffusion and reception of innovation. Rather than following a steady linear pattern, they tend to progress in leaps and bounds as the numbers of businesses adopting the innovation swell or shrink, servicing an expanding or contracting proportion of the market. As they grab an ever larger share of the market, other businesses get left behind. For them there is no point to investing in the innovation, since they have missed the chance to profit from it. At this point, there are two possible outcomes. They either go under or manage to ride the next wave of innovations. A typical example of the first outcome is the development of the American automobile industry (see Figure 5.1).

Box 5.3
Fuji: The one that looked ahead

'Fuji looked ahead and won,' read the headline. The article explained why the iconic Kodak is almost bankrupt while its one-time Japanese archrival Fuji is thriving. Once dominant on the photographic film market, Kodak watched its consumers defect to digital photography, leaving rolls of film on the shelf. Confronted with the same problem, Fuji decided to limit the production of photographic film to one location. It did all it could to squeeze as much profit as possible out of the old technology and invested in researching new products, a decision not made lightly. In the words of Peter Struik, MD of Fujifilm Europe based in Tilburg in the Netherlands, 'No one wants to kill the goose that lays the golden eggs.' What Fuji did was focus on cosmetics and medicine instead. This may seem like a weird decision, but its know-how in molecular technology and coating made the switch possible. Henk Volberda, Professor of Strategic Management and Business Policy at Erasmus University in Rotterdam, says the explanation for Fuji's success and Kodak's demise lies in Kodak's Anglo-Saxon business model. It was mainly aimed at its stakeholders and had a short-term perspective. Fuji's strategy was much more long-term, and via targeted staff transfers it provided room for creativity.

Source: *NRC Handelsblad*, 12 January 2012

5.3 Conclusion

Schumpeter distinguished five types of innovation:

- *Process*: new production processes such as the use of robots;
- *Product*: new products such as the iPhone;
- *Organization*: new organizational models for business and industry such as the mobile office made possible by the laptop and cell phone;
- *Market*: new target groups or markets such as the young Chinese population;
- *Input*: new raw materials such as soya for biofuel.

Innovation requires courage and investments, so not every entrepreneur can try his hand at it. It also creates unrest, disturbing the balance on the market. In short, innovations have a selective effect. The businesses that are the first and the best to apply an innovation have a temporary head start and a monopoly. Innovations can push existing businesses over the edge if there is no longer room for them on the market due to the innovation (*creative destruction*). In short, in Schumpeter's vision and evolutionary economic geography, everything should be viewed from a dynamic perspective. Incremental and radical change are both seen as unavoidable.

The evolutionary approach has really taken off in economic geography, mainly as a result of mounting dissatisfaction with the drawbacks of classical and neo-classical economics, and the fact that it better reflects economic continuity and change. As in nature, the key to survival is the ability to adapt. In terms of industrial location choice, this means we need to devote far more attention to a business' starting position. Many new businesses are actually spin-offs. Successful microchip manufacturer ASML, a Philips spin-off, is still based in Eindhoven, the home of Philips. In light of this spatial continuity, it may be more useful for policy-makers to focus on boosting the regional economy than to try to attract new businesses (see Chapter 9).

6

Classical regional growth theory: The importance of measurability

6.1 Introduction

Why is one country more economically successful than another? Why do levels of economic wealth sometimes vary greatly from region to region within a country? Economic geography explains these variations in development. Besides being of academic interest, national and regional comparisons serve a political purpose. Whether politicians work at the local, regional or national level, the economic development of their territory is high on their agenda.

It is not easy to explain the economic development of regions. It is always tricky to distinguish cause and effect. Does economic growth come from within or does international trade allow a region's economy to develop? These questions are typically addressed by growth theories seeking to explain the economic development of regions and their positions vis a vis each other.

This chapter begins by discussing mercantilism as a way of improving national prosperity. We address the classical economic theories of Adam Smith and David Ricardo, look at Michael Porter's competitive forces approach, and finish with the annual reports on the competitiveness of nations, the *World Competitiveness Yearbook* and *Global Competitiveness Report*.

6.2 Mercantilism: The colonial gold rush

From 1500 to 1800, the various regions of Europe crystallized into nation states where the interests of politicians and merchants began to run parallel. Merchants paid taxes to the young nation states, and the money was used to develop governmental control systems and armies to protect the economic interests of the merchants and support their growth. This was achieved by monopolizing international trade routes and waging wars against rival nation states. The economic nationalism prevalent at the time is known as *mercantilism*. Gold and silver were valued highly and served as currency and financial reserves. The nations with the largest stores of gold and silver were the wealthiest and most powerful. To enlarge the supply, the nations needed to build up a trade surplus, increasing their gold reserves and ultimately their wealth. Colonialism was a typical example of mercantilist politics. Colonies were used to source cheap raw materials and sell expensive finished products from the colonial motherland. A good example was the 16th-century Spanish empire, founded on a monopoly on trade with large parts of Latin America and enabling the Spanish kings to accumulate enormous gold reserves. Mercantilism was based on political convictions, not academic theories.

6.3 Adam Smith, the father of classical economics

Figure 6.1
Adam Smith (1723-1790)
Source: en.wikipedia.org

The aim of mercantilism was to combine exports with the highest possible level of self-sufficiency. Exports yielded gold and silver and imports were paid for by gold and silver. The global trade system was viewed as a zero sum game. One country's loss was another country's gain, and by definition, the sum of all the gains and losses was zero.

In the Age of Enlightenment (ca 1650-1800), a more optimistic, less belligerent worldview began to take hold. The Scot Adam Smith, who laid the foundation for classical economics, did not agree that a country's wealth was determined by its gold and silver reserves. 'The division of labour is the great cause of the increase of public opulence, which is always proportioned to the industry of the people, and not to the quantity of gold and silver as is foolishly imagined' (Smith 1904, p. 15). According to him, genuine wealth was not about gold and silver but about how labour was organized. He advocated labour division, enabling workers to become more skilled and more efficient. If two countries engaged in a cost-effective exchange of specialist goods or services, they would both become wealthier as a result.

Smith described two advantages that led to cost-effectiveness. The first was a 'natural' or 'absolute' advantage, the availability of a scarce natural resource or a particular climate or geographic feature. Roses thrive in warm regions without sharp temperature fluctuations, making equatorial regions like Kenya and Ecuador particularly suitable for growing them. Coal is easiest to mine from an open-pit mine, as in the American state of Wyoming or the Colombian *Departamento del Cesar*. The second advantage was an 'acquired' one. Producing goods or services required particular knowledge and skills. In Smith's days, the English textile industry was the most efficient in the world. Trade in grain and textile between England and the American colonies was beneficial to both. In Smith's view, there was no such thing as a zero sum game.

6.3.1 Ricardo's theory of comparative advantage

Classical economist David Ricardo (1772-1823) elaborated upon Smith's ideas on the value and advantages of regional specialization, adding 'comparative' advantages to the absolute ones.

Table 6.1
An example of absolute comparative advantages

	Country A	Country B
Costs of product X	10	20
Costs of product Y	20	10

In the example in Table 6.1, product X costs twice as much in country B as in country A. The reverse applies to product Y. However, it is not at all certain that buyers (consumers, businesses) in B are willing to buy X from A or that buyers in A are interested in buying Y from B. It depends on the *terms of trade*: the quantity of goods bought (imported) and the quantity of goods sold (exported). This ratio is shown in Table 6.2.

Table 6.2
An example of national terms of trade (two products, two countries)

	Country A	Country B
Costs of product X	1X = 1/2 Y	1X = 2 Y
Costs of product Y	1Y = 2 X	1Y = 1/2 X

Rather than purchasing X on the domestic market, residents of B buy it in A, where it only costs ½ Y – four times cheaper than in B, where it costs 2 Y. To meet the demand from B, A will have to expand the production of X. This requires extra factors of production. Since they are limited, A can only produce more X by producing less Y. As A can buy Y far more cheaply in B, it does not make much sense for A to produce Y itself. A comparable shift in factors of production from X to Y occurs in B. The two nations both specialize, A in X and B in Y, and both benefit from this. Thanks to international trade, consumers can either spend less than if they only bought their goods domestically, or they can buy more for the same amount of money. Table 6.3 gives an example of how Ricardo's law of comparative advantages functions.

Table 6.3
An example of relative comparative advantages

	Country A	Country B
Costs of product X	10	15
Costs of product Y	20	25

In this example, although X and product Y are both far more costly in B than A, trade is still an attractive option. Table 6.4 shows the terms of trade.

Table 6.4
An example of national terms of trade (two products, two countries)

	Country A	Country B
Costs of product X	1X = 0,50 Y	1X = 0,60 Y
Costs of product Y	1Y = 2,00 X	1Y = 1,67 X

Despite the high price levels in B, it is still sufficiently advantageous for A to buy Y there. The value of Y in A is 2 X, whereas in B it can be purchased for 1.67 X. Here as well, there is trade with A, which specializes in X, and trade with B, which specializes in Y.

This theory of relative comparative advantages demonstrates that differences in productivity rather than prices determine the scope of international trade. Proponents of the classical and neoclassical theory of comparative advantages hold that regions should specialize in the products or services they are clearly good at producing.

An important presupposition in the work of classical economists like Adam Smith and David Ricardo is market stability, a long-term state of balance between supply and demand. All the requirements are met and there is neither under- nor overproduction. According to Smith, stability is the result of an 'invisible hand,' his metaphor for the interplay between self-interest, market forces and the law of supply and demand. Together, the three mechanisms create economic stability. If the demand for Y rises, the price goes up and more of it is produced. If the demand for X falls, the price goes down, resulting in a fall in production or even a discontinuation of the product. The economy thus has the ability to self-regulate by adjusting production levels. This same mechanism applies to imbalances between nations and regions.

There are several assumptions in Smith and Ricardo's theories that deviate from reality. The theory of absolute and comparative advantages assumes countries can reallocate their factors of production to other products without any extra costs. The theory also focuses on the exchange of products at the expense of services. Classical regional trade theory does not allow for transport costs, trade barriers or differences in technological development between nations or regions. Moreover, it only recognizes one type of market, i.e. perfect competition.

Smith and Ricardo were keenly aware that the invisible hand did not always work. Smith was surprised at the lack of labour mobility between London and its surrounding regions at the time despite a wage difference of 25%. 'Such a difference of prices, which it seems is not always sufficient to transport a man from one parish to another. It appears evidently from experience that a man is of all sorts of luggage the most difficult to be transported' (Smith 1904, p. 30).

6.4 Basic and non-basic economic activity

Adam Smith concluded that the growth of cities depended on the buying power in the cities and the surrounding regions. The only way for an urban economy to grow, he thought, was via the spatial expansion of the market by exporting to a region further away. In his view, the construction and expansion of road networks to other regions and countries were vital prerequisites to growth.

Building on Smith's ideas, German economist Werner Sombart (1863-1941) drew a distinction between *basic* and *non-basic economic activity*. Basic economic activity means exporting products from a region, and non-basic activities means supporting the basic industries and providing goods and services for regional consumers.

Basic activities bring in foreign currency and expand a region's buying power. The growth of a region is the sum of the combined growth of the basic and non-basic industries, although the non-basic sector does depend on the basic sector. For a village to have a closed economy, it cannot just have bakers because there would not be enough villagers to buy all the bread. The non-basic sector needs basic activity. An example of basic activity is mussel farming in the village of Vrouwenpolder in the southwestern Dutch province of Zeeland, where there is an abundance of water. Many of the mussels farmed there are sold to Belgium. This allows the mussel farmers to employ more personnel than if they were to produce solely for the domestic market. The personnel spend part of their wages at the local baker's and hairdresser's, helping to ensure via their buying power that the non-basic businesses make enough money to remain viable. According to Sombart's export base theory, regional growth ultimately depends on a region's basic economic activity. In the Dutch village of Vrouwenpolder, any growth is thanks to the mussel farm, not the baker or hairdresser.

In applying the export base theory, one thing should be kept in mind. Smaller regions are more externally oriented and by definition have a more open economy. If a mussel farm sells its produce to a restaurant in Bruges, this is export, even if Bruges is only 50 kilometres away. If Cisco Systems in San José, California develops a network system for Harvard University in Boston, this is not export, even if the buyer and seller are 4,300 kilometres apart.

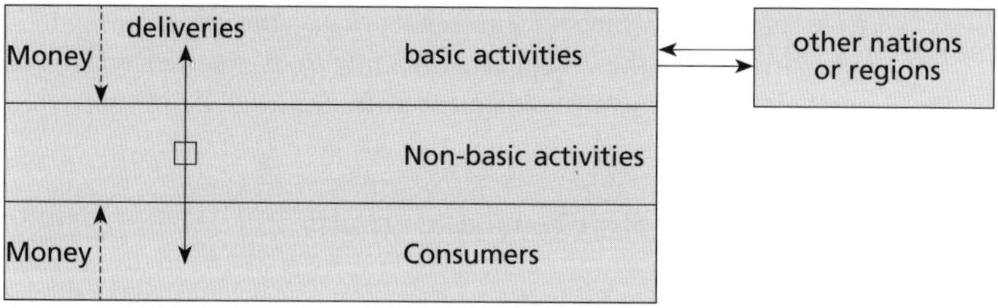

Figure 6.2
Export base theory

The distinction between basic and non-basic economic activities can be an arbitrary one. A cinema can serve a regionally basic function, drawing an audience from the nearby villages. But if the cinema is scaled up to serve a province or county, it is more like a non-basic service since not many people leave their own local district to see a movie.

For all its limitations, export base theory is still frequently used in decision-making, for example on which companies qualify for investment funding from a regional development organization or temporary support for businesses facing a drop in demand. In 2011, when the horticultural sector was struck by EHEC bacteria, the European Union set up a financial aid programme to keep this basic sector afloat. In the Netherlands, the number of butchers fell by 40% from 2000 to 2010, but no matter how much the meat lobby supports butchers, the process is not likely to be reversed. Job opportunities provided by independent butchers are disappearing and supermarkets can only partially compensate for this (like butchers, supermarkets play a non-basic role). In our discussion of *multiplier* and *growth pole theory* in Section 8.3, we return to the distinction between basic and non-basic economic activities.

6.5 Classical and neoclassical growth theory: Criticism and adjustments

The classical and neoclassical focus on equilibrium is still an important source of inspiration for politicians, international organizations, consultants and economists. To this day, the International Monetary Fund, *The Economist*, the World Trade Organization and liberalist political parties can all trace their worldviews back to the ideas of Adam Smith and David Ricardo. Wherever there is free trade, capital and products, weak regions can use stronger ones to boost their development. The arrival of Bulgarian labour migrants in the Netherlands was beneficial to Dutch employers, who now had motivated workers willing to do unpopular jobs, and to the migrants' country of origin, since their families now had more buying power.

Yet this concern with equilibrium has also been criticized. Why do the northern Dutch provinces lag behind the rest of the Netherlands economically? Will the southern Italian region of Cantabria ever be on an equal footing with northern Lombardy? One of the main critics of classical and neoclassical economics was British economist John Maynard Keynes (1883-1946). He doubted the effectiveness of the 'invisible' hand, wondering if it might need a helping hand itself at times. Much of the rampant unemployment in the Western world of the 1920s and '30s, he observed, was structural. The price mechanism was not functioning well enough to restore the balance between the labour supply and demand. Wages were protected by collective bargaining agreements so it was not possible to lower them. To Keynes, wages were not just costs incurred by employers. If a company paid lower wages, he reasoned, the workers would have less to spend and businesses would sell less. This is why, Keynes argued, businesses looking to invest were more interested in regional variations in profits than variations in wages. And profits depend on the demand on regional markets. Keynes also saw a strong link between gross regional product and regional employment. Economically deprived regions tend to have high unemployment rates. The solution, he stated, is to boost employment by raising the gross regional product, and he felt the best way to achieve this is by stimulating the 'effective demand' – spending incentives based on consumption, investment, state spending and the export of products. Keynes thought it was up to the state to lead the way in this connection, e.g. by hiring the unemployed for large infrastructural projects. Thanks to state job initiatives, people otherwise unemployed would be able to spend again. This is known as the *multiplier effect*. Once the economy is back on an even keel, the state can abandon the policy. In the Netherlands, examples of Keynesian economics include the construction of Eelde Airport and enormous parks near Amsterdam, Nijmegen and Rotterdam, large-scale infrastructural projects the unemployed had little choice but to work on.

The post-World War II era was not only a period of redevelopment, it was an era of increasing geographic decentralization. In the Netherlands, rural employment fell because of the increased mechanization of the agricultural sector. Industrial growth was mainly concentrated in the urban centres so there was an imbalance in the labour supply and demand. In 1951, 42 municipalities across the mostly rural northern, southern and eastern regions of the country were given 'industrial growth centre' status. Industrial companies were eligible for a 25% subsidy on their building costs in any of these 42 communities (see Box 6.1).

Box 6.1
Decentralization of government institutions in the Netherlands

In 1967, the Dutch authorities proposed plans to decentralize employment. It was predicted that the country would have twenty million inhabitants by 2000, so certain state institutions would have to relocate to the northern and southern provinces of Groningen and Limburg. Initially the plan involved fifteen institutions including Netherlands Statistics and the main pension fund. In 1972, the state decided that 20 to 50,000 jobs would be moved to these peripheral regions.
In practice, however, relocating was not such a simple matter. When it was announced in 1974 that the headquarters of the PTT, the national postal and telecommunication service at the time, would be moved to Groningen along with 4300 jobs, mass protests broke out among its employees. They did not like the idea of getting 'deported' to the far north, and people were afraid the PTT would lose valuable talented workers. In June 1975, in the face of strong resistance from employees as well as directors of the largest employer in the country, the Dutch House of Representatives toned down its decision. By the end of 1978, a mere 287 workers had moved from The Hague to Groningen. A total of 2200 rather than 4300 PTT jobs would move to Groningen and another 1000 to Leeuwarden in the northwest. The PTT was privatized in 1989, even before the new office adjacent to Groningen train station opened in 1990. By then 1400 workers had moved from The Hague, but the new middle management did not care for the decentraliza-

tion. There they were in Groningen while the top-level management of the postal and telecom divisions were still in The Hague. In March 1985, former Prime Minister Ruud Lubbers scrapped the policy of decentralizing employment, calling it an 'impassable dead-end road.' The plans of former Labour MP Wierenga to turn Groningen into a 'sub-seat' of government after The Hague were similarly shelved.
Source: *Reformatorisch Dagblad* 1990

The idea of decentralization also caught on at the European level. At the signing of the Treaty of Rome in 1957, the six member states of the new European Economic Community (France, West Germany, Italy, the Netherlands, Belgium and Luxembourg) announced their intention to strive for economic unity, reduce regional variation in economic growth and support economically deprived regions. The European Regional Development Fund was established in 1975 with this in mind. The European Commission used the fund to co-finance investments in motorways, airports, marinas, water purification plants, bridges and the restoration of derelict urban areas. The underlying thought was that the investments would stimulate regional economic growth and employment and help reduce regional differences. From 2007 to 2012, the fund had approximately thirty billion euros at its disposal.

6.6 A revival of classical economics: Porter's competitive forces theory

The 1973 international oil crisis triggered a prolonged period of economic stagnation and widespread unemployment in the Western world. In Rotterdam, the shipbuilding firm Rijn-Schelde-Verolme (RSV) folded in 1983, even though it had recently received government funding to the tune of what would now be a billion euros. In the second half of the 1980s, partly due to the RSV affair, economic state intervention came to be seen in a different light. The policy of pumping large amounts of cash into the economy had driven up budget deficits and inflation, and the time had come for a drastic reassessment of the whole issue of regional economic development.

In London and Washington, Prime Minister Thatcher and President Reagan were firm believers in classical economics. Only the free market could restore the economic balance. Slashing government spending to boost public finances, curbing the power of the trade unions and privatizing state institutions were seen as the way forward for the national and regional economies. Imbalances between regions were considered less important, partly because the regions traditionally the economic powerhouses were now in trouble themselves. In the Netherlands, the 1988Planning Policy Guidance Note 4 replaced the decentralization policy with the notion that regions were responsible for their own economic growth. This opened the door to a revival of classical economics.

It was American business strategy expert Michael Porter who took up the gauntlet. Towards the end of the 1980s, he launched a new perspective on regional economic development. In 1990, after four years of research he published *The Competitive Advantage of Nations* (see Chapter 1). Having noted that successful companies in a particular industry tend to come from the same region, Porter began his research by conducting an empirical analysis of the United Nations world trade statistics. He initially focused on various countries' shares in world export. In 1985 for example, Switzerland's share was 1.6%. If the market share of a product group rose above 1.6%, it could be an indication of strong competition. In Switzerland, this happened with watches (34.1%) and herbicides (20.6%). In the Netherlands in 1990 and 2007, a team of researchers led by Danny Jacobs followed Porter's research method. Export figures showed that the Dutch had a strong position in the agro-food sector.

Table 6.5
Shares of product groups in world export

	share in world export (%)		size of Dutch export (million $)	
product group	**1994/95**	**2004/05**	**1994/95**	**2004/05**
cut flowers	67	59	2140	3116
mussels	58	47	74	128
seed potatoes	67	57	312	258
potatoes	18	11	261	171
cocoa butter	42	33	561	811
cocoa powder	55	38	191	498
suckling pigs	51	29	142	178
pigs	37	23	423	396
vaccines	27	29	120	450
beer	22	19	834	1551

Source: Uncomtrade

Porter's analyses of the competitive forces of nations work from a classical economic view of national economies. A country's export position is an important indicator of its economic vitality. Switzerland is noted for its watches and pharmaceuticals, Germany for its engineering and luxury cars, and Italy sets the standard for ceramic tiles.

Had David Ricardo lived in the modern age, he would have reached the same conclusion. Porter however went beyond Ricardo's ideas, using his interpretations of trade statistics as the basis for his theory of competitive forces.

He added the conceptual 'diamond model,' which he used to take a closer look at industries with a relatively large share in world trade. He asked questions like: How do these companies interact? Do they have their own research facilities? What is the role of the state in these sectors? We discuss Porter's theory of competitive forces again in Chapter 8 when we examine the institutional approach to economic geography.

In the Netherlands, one of the conclusions drawn on the basis of the Porter analysis is that Dutch competitiveness is largely determined by the country's favourable geographic location in relation to the European market. Logistics and transport are two of the pillars of the Dutch economy, and the agro-industry is a third. If we apply Porter's methodology, we see that cut flowers, seed potatoes and pigs are highly competitive product groups.

Regional economic development is a key field of interest in economic geography. A region's economic dynamics are ultimately determined by what they have to offer commercially. As is noted in Chapter 2, the business climate can be analysed at different spatial levels:

- *The macro-level.* Some economic parameters are uniform within a country. Wages are determined via collective agreements that apply across the board. Sales tax percentages for different product groups vary between countries, as do borrowing interest rates and administrative obligations for companies.
- *The meso-level.* Some location factors have a particular regional relevance. An international airport has a wide regional reach. Labour markets for unskilled workers tend to be regional, and EU funding, e.g. from the European Social Fund, is determined at the regional level.
- *The micro-leve.* Some economic geographic location factors are highly local. Can people get to work by public transport? Will a business looking to set up shop in a residential area get planning permission? What message do the premises convey at a particular location?

Some economic parameters translate directly into commercial location factors. The regional economic parameter 'labour costs' can be measured by looking at the minimum wages levels in different countries. The capacity of a seaport can be measured by its 'accessibility by sea.' It gets more complicated if location factors are less unequivocal and have to be compiled from various regional economic parameters. So in addition to labour costs, anyone interested in labour productivity should also look at the workers' level of education, annual leave allowance and absence due to sickness. In the example of the seaport, accessibility can be as much a matter of how rapidly goods are cleared through Customs as how many container ships it can accommodate. Not all location factors that play a role in regional economic development can be expressed as figures. How do you express the quality of the living environment or the fellow entrepreneurs' willingness to collaborate? How do you weigh these factors against the price of electricity or rules regarding dismissal? Large businesses often call in a consultancy firm to compare the available locations before choosing a country, region and location.

Considering the strong position of the Netherlands in agro-foods, trade, distribution and bulk chemicals, it stands to reason that the country is more successful at attracting foreign distribution companies than the high-tech industry. The Dutch 'diamond' is more suited to the location factors required by the transport and logistics sector (Figure 6.3).

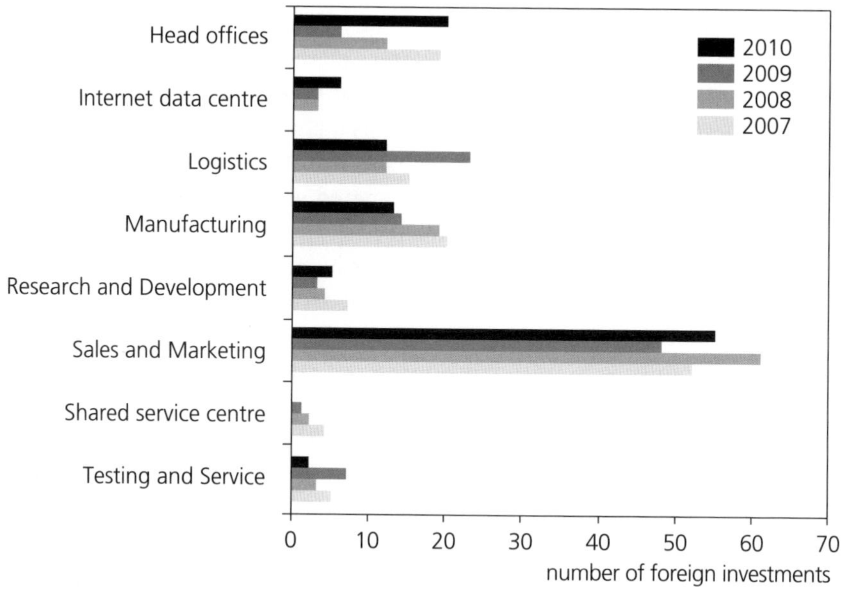

Figure 6.3
Foreign investment in the Netherlands by type of activity
Source: Ernst & Young 2011

Politicians want to know how to enhance the business climate in their regions and attract successful companies, and businesses need to be able to compare regions and make a choice. These 'economic agents' all need information and advice to be able to make informed choices. Where their needs intersect, a new business sector arose about thirty years ago, offering research and location advisory services. They advise businesses on the best locations and local government on how to make their districts more attractive to businesses. Examples of companies providing these services are STEC Group, Buck Consultants International and Ernst & Young.

6.7 Comparing nations and regions

British physicist Lord Kelvin once said, 'If you cannot measure it, you cannot improve it.' With so many companies operating on a global scale, a country or region's business climate is more important than ever. As a result, there is a growing need for reference material. Is the commercial climate more attractive to logistics companies in the Netherlands than in Belgium? Was the American Intelligent Community Forum's 2011 decision to award the Dutch town of Eindhoven the status of the world's *Intelligent Community of the Year* justified? In 1979, the World Economic Forum (WEF) announced its intention to measure countries' competitiveness. The aim of this international think tank founded in 1971 is to stimulate the world economy. At its annual meeting in Davos, Switzerland, politicians, entrepreneurs and academics exchange ideas about the way forward. Every year in September, the WEF publishes a *Global Competitiveness Report*. The 2011 edition compared 142 countries on the basis of 113 statistical indicators such as the quality of the infrastructure and the educational level of the population. To make the indicators comparable, multi-criteria decision analysis is used, a statistical technique for standardizing different types of indicators. Of the 113 variables, 78 derive from surveys conducted in each of the countries. The remaining variables are collected through international institutions such as the IMF and the World Bank.

Until 1988, the WEF published the *World Competitiveness Yearbook* in conjunction with the Swiss International Institute for Management Development (IMD) in Lausanne (table 6.7). In 1989, following a disagreement about the best way to measure competitiveness, the two institutions went their separate ways. The WEF sets more store by how entrepreneurs judge the national economies (78 indicators based on research into perceptions, 35 macro figures), and the IMD places more value on hard figures (116 perception indicators, 132 hard data). Moreover, the combined hard figures weigh more heavily (2/3) for the IMD than the figures from the perception research (1/3).

The WEF maintains that the growth potential of the GDP per capita determines a nation's competitiveness. The IMD is more focussed on the quality of the business environment. Despite these differences, the WEF and the IMD ranked the same 16 countries in the top 20 for 2011. In the WEF's *Global Competitiveness Report*, the Netherlands was in 7th place, but in the *World Competitiveness Yearbook* it fell from 12th to 14th place on the list of 59 countries (table 6.6). According to the IMD, in 2011 the business environment in the Netherlands was adversely affected by a poor economic climate in terms of youth unemployment, a drop in foreign investments and less interest in social cohesion.

Table 6.6
Top 20 of the World Competitiveness Yearbook and the Global Competitiveness Report

World Competitiveness Yearbook (2011) (59 countries)		*Global Competitiveness Report* (2011) (142 countries)	
1 Hong Kong	11 Luxembourg	1 Switzerland	11 Hong Kong
2 Singapore	12 Denmark	2 Singapore	12 Canada
3 United States	13 Norway	3 Sweden	13 Taiwan
4 Sweden	14 Netherlands	4 Finland	14 Qatar
5 Switzerland	15 Finland	5 United States	15 Belgium
6 Taiwan	16 Malaysia	6 Germany	16 Norway
7 Canada	17 Israel	7 Netherlands	17 Saudi Arabia
8 Qatar	18 Austria	8 Denmark	18 France
9 Australia	19 China	9 Japan	19 Austria
10 Germany	20 Great Britain	10 Great Britain	20 Australia

In both lists, the Netherlands ranks both high and low (see Table 6.7).

For regional politicians and location advisors, comparisons like this are too low-resolution and some indicators are best measured at the meso- or micro-level anyway. Following the success of these national rankings in terms of publicity and policy development, recent decades have witnessed a proliferation of regional rankings (see Table 6.8). This was made possible by the improved availability of regional comparative statistics, e.g. as a result of the work of Eurostat. In 2010, the European Commission's Joint Research Centre developed the EU Regional Competitiveness Index (ERCI), which compares 268 European regions at NUTS 2 level (provinces in Belgium and the Netherlands, Regierungsbezirke in Germany, counties/unitary authorities in the UK). With 68 variables, there are fewer variables than at the national level. One of the employees of a Dutch regional development organization tasked with attracting foreign investors drew attention to the relative importance of the competitiveness rankings. 'We mainly used our high ranking to influence the media; I don't know a single entrepreneur who came to us because of our high ranking. Having said that: this sort of information may get stored in an entrepreneur's subconscious. If he finds himself looking for a new location a few years later, he may end up with us after all.'

Table 6.7

Highest and lowest rankings for the Netherlands in 2011 (based on WEF and IMD data)

Italicized text indicates data were obtained through surveys among entrepreneurs.

Position according to IMD (out of 59 countries)		**Position according to WEF** (out of 142 countries)	
1	*National culture is open to foreigners*	1	Malaria infections per 100,000 inhabitants
1	Women in the workforce	1	Annual inflation
1	Dismissal costs	2	Quality of seaport infrastructure
2	Infrastructure for shipping	2	Secondary school attendance
3	Available medical care	2	Competition policy
3	Broadband take-up	2	Broadband take-up
3	Knowledge of foreign languages	4	Trade tariffs
3	Broadband speed	4	Quality of marketing of enterprises
5	Foreign direct investment	5	Collaboration between employers and employees
46	Costs of mobile telephone network	66	Relation between labour costs and productivity
46	Labour costs of managers	68	Solvency of banks
47	Labour costs in the service sector	72	Overall tax burden
48	Costs of broadband	93	Protection of investors
50	Working hours	97	Government deficit as % of GDP
55	Income tax as part of per capita GDP	113	Government debt as % of GDP
58	Incoming investments as % GNP	121	Government regulation regarding dismissal
58	Threat of relocation of R&D facilities	132	Flexibility in wage determination

Table 6.8

Top 20 EU Regional Competitiveness Index 2010

1	Utrecht (NL province)	11	Oberbayern (Upper Bavaria)
2	Copenhagen and surrounding area	12	Surrey, East and West Sussex
3	North Holland (NL province)	13	Gelderland (NL province)
4	London	14	Gloucestershire, Wiltshire
5	Stockholm	15	Darmstadt
6	Helsinki and surrounding area	16	Limburg (NL province)
7	South Holland (NL province)	17	Brussels
8	Paris and surrounding area	18	Bedfordshire, Hertfordshire
9	North Brabant (NL province)	19	Vienna
10	Oxford and surrounding area	20	Hamburg

6.8 A critical look at the competitiveness index

The concept of the competitiveness index has been criticized by various authors. The most prominent among them is Economics Nobel Prize winner Paul Krugman, who was adamant that countries do not compete with each other in the same way corporations do. Companies tend to grow at the expense of their competitors, but economic growth in one country can be highly beneficial to growth in another. He called politicians' obsession with competitiveness dangerous, saying it arises from nationalism and can lead to a call for privileges for strong industrial sectors and to protectionism.

Sometimes the rankings do not allow for any relation between certain variables. The Dutch did not do so well with the IMD in terms of working hours, but scored well on women in the workforce. Compared with the international norm for women in the workforce (a minimum of an hour a week), the participation of women in the Netherlands in 2010 was very high (72%). In Europe, only Denmark scored higher with 74%. Nowhere else in the EU did as many women work part-time that year (74%) as in the Netherlands. Germany was second with 45%. Whether Dutch women had an easier time making the arrangements needed to enable them to work part-time was beyond the scope of the analysis.

Another indicator that is hard to interpret is the tax burden. Though they were viewed as negative in the listings, it may well be thanks to high taxes that Scandinavian countries were able to develop a strong innovation profile. Through its investments, the Finnish Funding Agency for Technology and Innovation Tekes enabled Finland to be one of the few European countries to comply with the Lisbon Treaty (3% of GNP to be spent on R&D). Perhaps how a country spends its taxes is more important than the actual tax levels.

A final word of criticism pertains to the data collection for the surveys conducted by the IMD and the WEF. The IMD had the data collected by alumni from its own Business School. In 2011 the compilers of the *World Competitiveness Yearbook* received 4935 survey results from 59 countries, an average of 48 per country. The WEF received 14,039, varying from 34 for Surinam to 422 for the U.S. So quantitatively and qualitatively, this is a relatively small sample and it would be tenuous to draw far-reaching conclusions from it. For example, in 1999 the holiday destination Mauritius with a population of 1.2 million scored higher than South Korea with a population of 48 million in terms of its research environment – a somewhat puzzling outcome (Van Hoof 2012).

However, Stéphane Garelli, Director of the IMD World Competitiveness Centre, claims the rankings are continually improving. He argues that the IMD index not only measures economic achievements, prosperity and power, it also reflects a prolonged period of research and the accumulation of knowledge and insights. Businesses compete and create wealth, but the context in which they do so is determined by governments. Garelli (2011) takes the example of East Asian nations. He stresses the importance of education and knowledge development for competitiveness. 'In recent years, the results highlight the formidable efforts that East Asian nations have made to improve education. In addition to being competitive (temporarily) because of cheap labour, they aim to develop their competitiveness level so that it is based (permanently) on an educated workforce. Knowledge is perhaps the most critical competitiveness factor. How that knowledge is acquired and managed is each nation's responsibility' (Garelli 2011, p. 491). For that reason alone, countries compete. It is not enough to look at nations' export positions at the level of goods or services. However, as Garelli points out, 'in the early 1980s the world had changed. In addition to trade, foreign direct investments (FDIs) had become a major force in world economics. Nations discovered that there was a new avenue to prosperity and this was precisely by attracting FDIs. This argument was subsequently vindicated by the remarkable success of Singapore, Ireland or Dubai, which based the early stage of their competitiveness on encouraging foreign investments.' Having said that, he cautions that 'as any business leader has experienced, a similar operation with identical products, technologies and processes can perform very differently in country X or Y' (Garelli 2011, p. 492).

6.9 Conclusion

Comparing the economic strength of nations is as old as international trade itself. In the late medieval period, the focus was on gold, and in the Industrial Revolution it was on absolute and comparative advantages. The 1990s witnessed the rise of competitive forces theory, enabling researchers to compare how regions and industries compete. From the aspects of business management, regional economics and economic geography, the study of the competitiveness of nations and regions grew into an empirical discipline in its own right, watched closely by policy-makers and politicians. The differences in the perspectives and insights of the World Economic Forum and the IMD Business School highlight one of the fundamental research questions in economic geography and economics regarding economic development. Is it the hard macro-economic figures that enable Adam Smith's 'invisible hand' to do its work and ensure that businesses 'automatically' settle in the best regions? Or is the vitality of a national or regional economy mainly determined by corporate decision-makers' perceptions of the economic situation in a country or region? In Chapter 7 we see how academics and policy-makers who believe in the power of perception have developed their own field of study: *place marketing.*

7

Place for sale: Attracting visitors, businessmen and residents

7.1 Introduction

It is clear in Chapter 6 how regions compete on the basis of 'hard facts' and 'soft perceptions.' Neoclassical location theory holds that entrepreneurs always act rationally and base their choice of location on hard figures, costs, the labour market and selling opportunities. It is the job of local, regional and national authorities to have the figures make their regions look as good as possible and boost their popularity. From this perspective, an entire field of research is now focused on comparing nations. This discipline has witnessed strong growth since 1979. However, nations are not only compared on the basis of facts and figures. The perception of entrepreneur may be subjective, but it plays a significant role in the decision-making. Research into how nations, regions and cities are perceived and the image they have is highly valued. British nation branding expert Simon Anholt makes policy recommendations to national, regional and local authorities and has developed the Nation Brands Index and City Brands Index.

The notion that perception can be influenced has given rise to a new field of policy studies: *place marketing*. This interdisciplinary discipline combines the expertise of economic geographers, psychologists, economists, marketing and business management experts and communication specialists.

7.2 Mass migration and day-trippers

The promotion of regions and specific places has a long history. Roman Emperor Julius Caesar tried to persuade 100,000 poor residents of Rome to leave the city and settle in new colonies all across the Roman Empire. In his term in office from 1906 to 1911, Russian Prime Minister Pjotr Stolypin had 6,000,000 flyers printed to encourage Russian farmers to resettle in Siberia.

In the Netherlands in the second half of the 19th century, the wealthy classes developed an interest in day trips to seaside resorts and the rolling hills of southern Limburg. On 22 February 1885, a regional tourist information agency as founded in Valkenburg, printed leaflets and placed benches in and around the town for day-trippers taking a walk. Affluent visitors were lured to the town by newspaper advertisements and murals on buildings. All the efforts of Julius Caesar, Stolypin and the first Dutch tourist information agency targeted potential settlers and tourists. Place marketing mostly consisted of advertisements and leaflets in the hope of attracting investors or tourists. As recently as 1982, Burgess wrote in *Regional Studies,* ' Image making could be improved via more creative and original copywriting, better market research and increased emphasis on public relations work.'

Figure 7.1
Early 20th-century advertisement for the English seaside resort of Blackpool
Source: Midland Railway Blackpool

7.3 Place marketing

In the late 1980s, Greg Ashworth and Henk Voogd argued that 'a marketing approach to urban planning is a very promising new avenue to explore' (Ashworth and Voogd 1988, p. 65). The reason, they said, is that the steady growth of competing urban activities (recreation, retail, housing) makes it worthwhile to carefully plan ways to reinforce the market position of cities. First used by American marketing expert Philip Kotler in 1993, place marketing is an umbrella term covering region branding, city branding and inward investment (see below). The main differences lie in the geographic scale (nations, regions, settlements), the target group (businesses, tourists, shoppers, present or future residents, international organizations) and range (enhancing the recognizability, attracting people and organizations).

By 1990, the rise of area marketing from a planning perspective was paralleled by an increasingly commercial state approach. In 1993, Kotler found that marketing principles applied to commercial products were also applicable to public services. In a world characterized by a global economy, cities, regions and nations are well advised to make good use of marketing techniques. One disadvantage compared with the commercial world, Kotler noted, is that they can't react as promptly to developments on the market. It is a lot easier for a soft drink producer to add a new flavour than for a city council to change the look of an entire city centre.

Kotler (1993) distinguished six things place marketing can do to enhance the market position of a city or region:

- attract recreational and business visitors
- encourage entrepreneurs to settle in the area ('inward investment management')
- keep and expand the businesses already there
- promote new enterprises and the growth of smaller businesses
- stimulate export and outgoing investments
- enable population growth or alter the composition of the population

Before local or regional policy-makers choose one or more of the approaches, they need to do their homework, often involving a SWOT analysis (Strengths, Weaknesses, Opportunities and Threats). What products and services are offered in the area that may be competing with each other? Where are the most promising markets? How can products and services be improved? How should the message be communicated to the target group?

In his book, Kotler gave checklists for how an area can surmount a crisis, suggesting that cities faced with economic and social decline revise their town planning. They should improve their infrastructure and public services such as education, fire and sanitation departments and develop attractions. An assessment should be made of the city's image, which should be accurate, credible, simple, attractive and distinctive. Ashworth and Voogd characterized cities' efforts to sell themselves as follows (Table 7.1):

Table 7.1
Strategic choices for place marketing

	Services and facilities	
Market	**Maintain**	**Adjust**
Present consumers	1 Consolidation	2 Quality
Future consumers	3 Expansion	4 Diversification

Source: Ashworth and Voogd 1988

The first strategy can maintain the existing level of services for the current residents. It can be an important strategy for areas with declining populations. The second strategy can pertain to areas where the infrastructure is outdated and city centre rejuvenation for example can make a shopping centre more attractive to shoppers. The third strategy targets new markets for existing facilities. In 2011, Irish budget airline Ryanair started flying to Eindhoven Airport. The regional tourist information offices used the new connection as an opportunity to encourage Ryanair passengers to stay and explore the region. The aim of the fourth strategy is to bring new services to new markets. Examples include a purpose-built fashion outlet shopping village and a historical dockyard, both in Lelystad, the capital of the *polde*r province of Flevoland.

Once the desired image has been selected, it is time to take it to market. Various communication tools can be employed such as advertising, presentations at trade fairs, hosting events and using the place as a film set. There was an almost prophetic note to Kotler's final remark, 'Places are in trouble now, or will be in the near future. The globalization of the world's economy and the accelerating pace of technological changes are two forces that require places to learn to compete. Places need to learn how to think more like businesses, developing products and discover potential markets' (Kotler *et al.* 1993, p. 346).

7.4 The role of geography in place marketing

Greg Ashworth and Mihalis Kavaratzis (2005) criticised the use of marketing techniques to sell places. 'Very few marketing specialists have given much thought to its application to places treated as products and if they do, they too easily assume that places are just spatially extended products that require little special attention as a consequence of their spatiality.' According to the same authors (2009), this is due to the 'sex appeal' of city marketing. Politicians and authorities like to create policies with catchy slogans. Professor of Place Marketing Gert Jan Hospers (2009) observed that the motivation behind place marketing is often a wish to 'put a place on the map,' the thinking being that 'our area has so much to offer, but hardly anyone is aware of it.'

At the start of the new millennium, the new sub-discipline of place marketing developed, referred to in academic literature as place branding. People develop feelings for a place and attach meaning to it through:

- the physical vicinity including the built environment;
- the way they use places;
- a variety of representations of a place in the media, films and literature.

Within these three categories, a distinction should be drawn between the experiences of people themselves and those handed down to them. Together they constitute a mental map (see Chapter 3) of a place. The aim of branding is to influence people's mental map and make it more positive.

Brands are often defined as 'all the perceptions and images representing a company, product or service.' Ashworth and Karavatzis defined a brand as a way of communicating (see Table 7.2).

Table 7.2
Branding as communication process

Brand identity	Brand positioning	Brand image
How *brand* owners want the *brand* to be experienced	The aspect of the product/ service that is aimed at a target group and can demonstrate its competitive advantage	How the *brand* is perceived by the target group

Source: Ashworth & Karavatzis 2009, p. 523

They argued that a place cannot be compared with a single product. A maker of energy drinks can claim his products contribute to the body's speedy recovery after exercising. A city however has far more types of users than an energy drink. Shoppers, home buyers, entrepreneurs, tourists, students etc., it has to seduce them all to come settle or visit. There are however parallels with corporate branding. The customer base of a company like Philips includes a wide range of target groups from housewives to electricians and teenagers. Yet with 'sense and simplicity' it has created an 'umbrella brand' encompassing the entire company. According to Ashworth & Karavatzis, a city can have multiple brands as long as they are anchored in the same set of norms and values. This is not always straightforward, as is demonstrated by the slogan 'Nothing can top Groningen,' one of the longest-running place marketing campaigns in the Netherlands.

Box 7.1
Can brands clash?

Geographers Piet Pellenbarg and Wim Meester, both from the University of Groningen, introduced behavioural economic geography in the Netherlands. In 1988, they investigated perceptions of the northernmost Dutch province of Groningen and its capital of the same name. More than twenty years later, Professor Pellenbarg reviewed this period. 'The Dutch saw the people of Groningen as dour.' They knew the landmark Martini Tower, but thought the city was smaller than it actually was. Pellenbarg and Meester joined forces with a research bureau and a media company to translate the research findings into a marketing campaign and in 1989 the media company came up with the slogan 'Nothing can top Groningen.' Every two years from 1989 to 2005, the University of Groningen Faculty of Spatial Sciences evaluated the effect of the campaign by conducting fifty telephone interviews in each province. In 1992, more than 60% of the respondents thought the city of Groningen had fewer than 150,000 residents. By 2000 this percentage had dropped to 44. After 2000, the campaign put less emphasis on the economy and the city itself, and the percentage rose to 60% in 2005. The research indicated that the province of Groningen's weakness was its image in terms of employment and accessibility of business networks. It scored better in terms of recreation, education, culture and landscape. The polls also showed that the image of the 'dour northerner' is fading. The academics' involvement in the evaluations ceased in 2005. In 2004 the province started using the slogan 'Nothing can top Groningen' to boost tourism. Pellenbarg had his doubts about the effectiveness of targeting entrepreneurs and tourists with the same slogan. He felt promoting the region as one offering space and tranquility ran counter to promoting it as an attractive place for businesses to set up shop. He cited the example of the southern province of Zeeland. 'It promoted itself with the slogan "Living and working in a green oasis." So make up your mind, do you want more tourism or more jobs? Personally I think the latter is more important.'

7.5 Place marketing in the Netherlands

Many branding efforts focus on visitors. Professor of Place Marketing Gert Jan Hospers does not back this approach. He draws a distinction between 'warm' and 'cold' marketing. Warm marketing tries to bind local residents and businesses to the city and cold marketing is aimed at attracting new residents and entrepreneurs. In Hospers' view, warm marketing is more important than cold. 'People generally move over short distances, 94% stay in the same region, 75% even in the same municipality. In 2011, one in three European regions had shrinking populations, and by 2050, some predictions say, the European population will have shrunk by 50 million.'

Martin Boisen (2011) drew attention to the often selective nature of place brands. They conform to the neoliberal doctrine that competitive forces can lead to a biased presentation (see Chapter 6): Highlight the strengths and hide the weaknesses. When Amsterdam wants to attract wealthy visitors, it uses imagery in its advertising focused on the historical city centre with its famous canals.

Table 7.3
Branding campaigns in the Netherlands, October 2011

Target group	Scale	Institution	Branding
International businesses	The Netherlands	Netherlands Foreign Investment Agency	*Holland, pioneers in international business*
Foreign tourists interested in design and architecture	The Netherlands	NBTC Holland Marketing	*Dutch DNA 2011*
Potential new residents and entre-preneurs	Province of Limburg	Province of Limburg and other institutions	*Limburg the bright site of life*
Conference delegates and organizers of business meetings	City of Maastricht	Maastricht Conven-tion Bureau	*Meet in Maastricht*
Current and future residents	Rotterdam South	Rotterdam City Council	*Zuid wil vooruit ('South wants to move forward')*

Yet there are exceptions to the rule that branding is only a matter of highlighting strengths. In an effort to boost the image of the relatively unpopular district of Rotterdam South, in 2011 city authorities coined the slogan 'Zuid wil vooruit' (South wants to move forward) (see Box 7.2).

Box 7.2
Zuid wil vooruit

For years the Rotterdam authorities felt the district of Rotterdam South topped 'the wrong sort of charts' with high unemployment and school drop-out rates almost un-Dutch in their scale and intensity. The local council hoped branding could help create viable, attractive and self-sufficient residential areas. Within the district, Pendrecht and Katendrecht are examples of areas that have bucked the trend and developed in a positive direction.
The local authorities are keen for the district's 'DNA' to be communicated to the outside world and branding is viewed as the best tool. In Katendrecht, it was done with the slogan 'Can you handle the Kaap?' The message was that living in Katendrecht is not for the faint-hearted. To live in a big city, you have to take the good with the bad.
Pendrecht launched a similar campaign and the slogan 'Pendrecht is on the right track' put the residents at the heart of it. Every month for over two years, a new 'local hero' was on the billboards to promote a particular community project. The

projects aim to make the area more fun to live in and more pleasant to look at and to make the residents prouder. It is this can-do mentality that characterizes Pendrecht and the marketing campaign invites residents to take part in local community activities.
Source: *Woonstad Rotterdam*

Place marketing may have taken off in terms of governance (three quarters of the Dutch municipalities now have a place marketing-related planning policy statement), but academically the discipline is still in its infancy. Marketing expert David Gertner (2011) investigated 212 publications on place marketing and place branding from 1993 to 2009. His conclusion was that they are mostly essayistic articles based on personal experiences and descriptive case studies. So far very little robust research has been conducted in this area.

7.6 The hunt for bohemians, biomedics and brainiacs

The Rise of the Creative Class, an influential book by Richard Florida, was published in 2002. Florida held that creativity is the most important economic driving force, and creative industries are the engine of an economy. According to him, authorities wishing to boost local economies should invest in education, research and development and an attractive living environment. This makes a place more attractive to the creative class, which supports small and medium-sized businesses and facilities that in turn will attract more creative people.

Cities with stagnating economies, Florida argued, should focus on the three T's, Talent, Tolerance and Technology, which act as magnets to businesses. There is ample debate about how to define the creative class and Florida himself counted 40 million creative professionals within the US alone. He divided them into three groups (see Table 7.4).

Table 7.4
The creative class as defined by Richard Florida

Super-creative core	Bohemians	Creative professionals
Computer programmers, mathematicians and engineers, medical and other social professionals, teachers and librarians, artists, media workers, designers and professional athletes	Commercial designers, sculptors, photographers, writers, painters, composers, musicians, singers, actors, choreographers, dancers and film directors	Managers, financial service providers, the legal sectors, health care and sales managers

Source: Atzema *et al.* 2005

To test the significance of the creative class to the economy, Florida used indices such as the Creativity Index and Bohemian Index (see Table 7.4). He found correlations between high scores on the various indices and economic growth. The main conditions for a creative urban climate are the three T's, Talent (a talented, highly educated and skilled population), Tolerance (a diverse community with a 'live and let live' mentality) and Technology (technological infrastructure needed to fuel an entrepreneurial culture). The purpose of the Creativity Index is to make it possible to compare cities. It is made up of four equally weighted components:

- the share of the creative industries in overall employment;
- the per capita number of patents;
- the presence of high-tech industry;
- the percentage of gay people in the overall population (an indication of tolerance).

Creative workers feel most at home in a culturally rich and tolerant social climate. Their hedonistic lifestyle generates a rich array of activities, livening up the urban environment and attracting a wide range of individuals and businesses. According to the Dutch newspaper *NRC Handelsblad*, the Amsterdam City Council invested nearly 40 million euros in incubator facilities where artists and creative-industry-related businesses can find relatively cheap work space. The idea behind it is that creative individuals are inspired by each other's presence. Rotterdam tried to attract creative residents by selling renovated subsidized housing in older areas to new creative professionals.

However, not everyone views attracting creative professionals as a source of economic growth. Some researchers argue that Florida's evidence is tenuous and his argumentation weak. Economist Edward Glaeser (2011) pointed out that the correlation between the percentage of highly educated personnel and economic growth is stronger than the one between the presence of gay people and bohemians and economic growth. Some definitions of creative class are so wide that 47% of the Dutch working population belongs to this class of professionals. Florida's ideas nonetheless continue to be very popular. By late 2010, his speaker fee had risen to $40,000, and his Richard Florida Creativity Group advises cities and nations worldwide.

7.7 Hallmark events: Culture and sports as marketing tools

It is clear at the start of this chapter how places compete for business, residents, visitors and facilities. Organizing cultural and sports events can help boost a city's attractiveness to residents and visitors alike. Some events are one-offs, others occur annually. Major events are known as hallmark events, 'Major one-time or recurring events of limited duration, developed primarily to enhance the awareness, appeal and profitability of a tourism destination in the short and/or long term. Such events rely on their uniqueness, status, and/or timely significance to create interest and attract attention' (Ritchie 1984, p. 2).

One use hallmark events can be put to is urban renewal. British geographer David Harvey has argued that mega-events are an important tool for cities to convey a certain image, enhance their status and catch the eye of the world. An example of a hallmark event was the 1851 Great Exhibition, precursor to the World Fair, which drew six million visitors to Crystal Palace in London's Hyde Park. The 2012 London Olympics were a more recent example.

Hallmark events have developed into a popular way to stimulate urban renewal. In 1993, urban sociologist Hartmut Häußermann predicted the 'festivalization' of city centres, with cities hosting expos, trade fairs, Olympic Games, world championships and film festivals. Places such as Potzdammer Platz in Berlin, St Peter's Square in Rome and Times Square in New York are increasingly becoming 'consumable products.' Via hallmark events, authorities and organizers aim to brand or rebrand places and create a communal focus for the local population. These events are used to free up resources for investments in infrastructure and public space.

In areas or urban districts in need of a social or economic boost, cultural or sports facilities can be part of urban renewal schemes. Examples are the City of Manchester Stadium built in 2002 for the Commonwealth Games in the Eastlands district of Manchester and the Guggenheim Museum in Bilbao, opened in 1997. This striking building grew into a landmark for a surrounding area previously known for its social problems. The museum was used to revitalize Bilbao and provide an economic stimulus for a city that had witnessed an industrial crisis. Attracting about a million visitors a year, the Guggenheim has become the city's 'postcard symbol,' and architects and town planners now speak of the 'Bilbao Effect' (see Box 7.3) when a town succeeds in redefining itself.

Whether hallmark events and flagship buildings really yield the necessary return on investment is something often left to advisory agencies to determine. Events and construction projects of this magnitude are frequently preceded by an *ex ante* study. The aim of the study is to assess and quantify the economic and social effects of the project, and its *ex post* counterpart evaluates the project afterwards.

Ex ante studies are often optimistic. Prior to the FIFA World Cup hosted by Germany in 2006, the German Soccer Association and the German Chambers of Commerce jointly estimated *ex ante* that the German economy could look forward to a cash injection of 3.5 billion euros and the creation of 60,000 new jobs.

Ex post evaluations are less unequivocal regarding the effects of hallmark events. In a 2011 article in *The Economic Journal*, economists Rose and Spiegel observed a positive correlation between a bid to host the Olympics or the World Cup and an export rise compared with non-bidders. By contrast, German economists Feddersen and Maennig (2010) noted that the West German cities that hosted the 1974 World Cup and the U.S. cities that hosted the 1994 Cup exhibited negative employment development compared with cities without matches of this kind. The *ex post* evaluation of the 2006 World Cup in Germany showed it mainly resulted in a temporary rise in employment consisting of 2000 hospitality jobs (Brenke and Wagner 2007, p. 26). The positive effects of extra tourists arriving for an event are partly undone by residents who leave the city because of the event or tourists who avoid the country rather than being drawn to it (Oldenboom 2006, see Figure 7.2).

A Visitors who combine visiting a place with going to the event (*casuals*);
B Visitors who cancel another planned outing to go to the event (*time switchers*);
C Potential visitors who prefer to avoid the event (avoiding the host city or postponing a visit);
D Visitors who would not have come at all if it had not been for the event;
E Residents who stay at home while the event is on;
F Residents who leave town while the event is on.

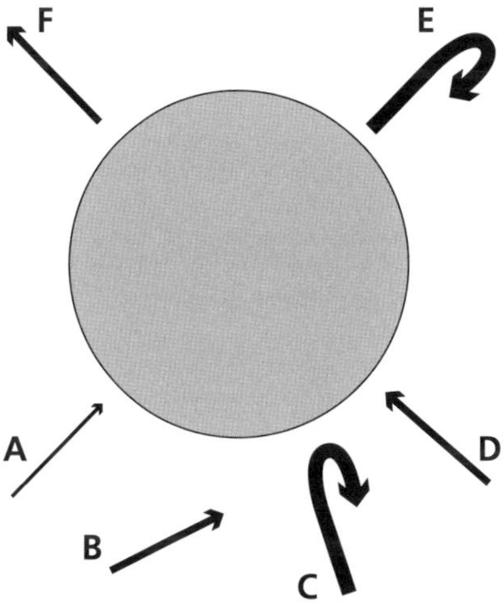

Figure 7.2

The effects of large sports events on visitor numbers to and from the sporting venue

Source: Oldenboom 2006

Categories D and E create a positive local economic effect, C and F a negative effect and A and B are neutral. This temporary reshuffling of the local population dynamics is often not referred to in *ex ante* studies but clearly influences the impact of the event. According to Feddersen and Maennig (2010), when judging the impact of an event, it is important to take into account the geographic range of the impact. The construction of the multi-purpose Max Schmeling Hall in Berlin increased the value of land in a radius of 300 metres around the building. Landmark buildings well embedded in the environment can have a positive effect on the surrounding area. However, Feddersen and Maennig argued that the most important effect of large sports events is the feel-good factor they create and the enhanced image of the country as a whole and the host cities in particular. Nation branding researcher Simon Anholt confirmed in his ongoing research that the perception of Germany improved after the Soccer World Cup.

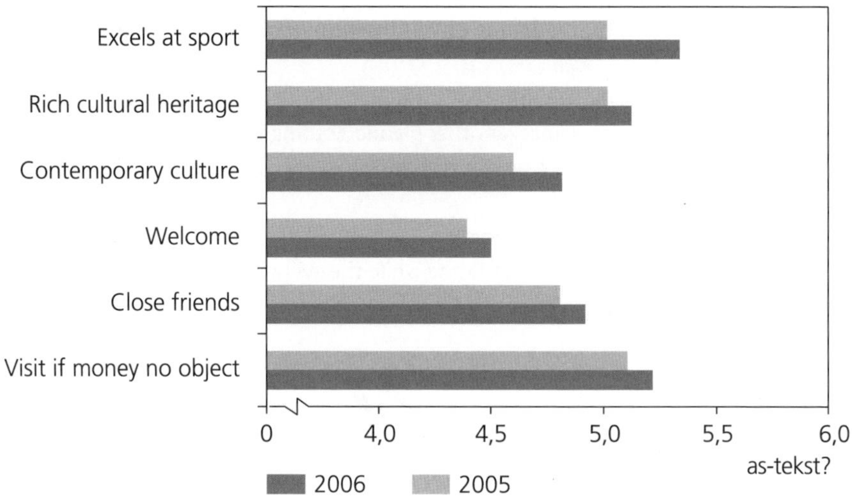

Figure 7.3

International perceptions of Germany before and after 2006 World Cup*

*On a scale of 1 to 7, 1 = completely disagree and 7 = completely agree.

Source: Maennig after Anholt, 2010

7.8 Conclusion

The promotion of tourism and emigration destinations is a form of place marketing that has been around for over a century. As of the mid-1980s, place marketing began to have more formal, organized applications, with authorities becoming increasingly market-oriented. Behavioural economic geography was among the first academic disciplines to develop theories and conduct empirical research into place marketing. Methodologies and terminology borrowed from consumer marketing gradually became the predominant investigative tools. Place marketing has attracted a great deal of interest from many other disciplines, and has proven relevant to policy- makers. The application of place marketing really took off in 1995. Whether this will turn place marketing into a fully fledged academic discipline remains to be seen. At the moment it is still very much in its infancy.

Box 7.3

Great expectations: the 'Bilbao Effect'

Towards the end of the 1980s, the Basque harbour town of Bilbao was in dire economic and social straits, with 25% of the population unemployed, the steel and shipbuilding industries in a slump and the city plagued by regular bombings by the Basque separatist movement ETA. Local and regional authorities decided the way out of the crisis was to transform the economy by shifting the focus from the primary sector to services. The idea was that investing in cultural projects would attract more tourists, which was no simple task in a city that thus far had little to offer visitors. The regional government made the bold move of asking the Guggenheim Foundation in New York to consider housing part of its modern art collection in a brand new museum to be built in Bilbao. The plan came under heavy fire. Trade unions felt the budget for constructing the building should be used to support existing industries. ETA later expressed its objections by attempting to bomb the building, which was narrowly avoided. The Guggenheim Foundation was more tempted by cities with an established cultural profile, such as Paris, Salzburg or Venice.

Fifteen years on, Bilbao is a different place. It attracts over a million tourists annually and has flight connections with about twenty European cities. The museum that triggered this transformation was designed by American architect Frank Gehry. The undulating, titanium-clad edifice is so photogenic it has received widespread attention in the international media. It was even used as a film location for the 1999 James Bond film *The World is Not Enough.* The museum was part of a wider urban regeneration programme also including a new conference centre, a metro and a new bridge across the Nervion River. From 1997 to 2003, researchers calculate, the Guggenheim Museum provided the city with an extra income of over 900 million euros. Additional tax revenues over the same period (143.5 million Euros) more than covered the 120 million euro building costs. Research suggests that 80% of the tourists come specifically to see the Guggenheim Museum, which has created some 4000 new jobs in the tourism sector.

Other cities have tried to copy the concept. The director of the Simon Wiesenthal Centre in Jerusalem wanted Gehry to design a museum there as well, explaining his choice of architect by saying the name Gehry alone would do the trick. You don't need to advertise, the building is advertisement enough. The building's unique design, he thought, would act as a magnet to the media. Besides Bilbao, the Guggenheim Foundation has museums in New York, Venice, Berlin and Abu Dhabi. In 2011 it announced it would be opening another museum that Taipei, Helsinki,

Guadalajara and Rio de Janeiro were all in the running for. The Foundation chose Helsinki and carried out a feasibility study in 2011 for a museum in the Finnish capital.
Sources: *Jerusalem Post* 2008; *The Guardian* 2007.

8

Institutions and intervention: Can spatial growth be directed?

8.1 Introduction

The Issue of whether governments should intervene in the economy has been the subject of a lengthy debate with different theorists setting the tone at different times. The Dutch are forever seeking a balance between 'let the market do its thing' and active intervention. For a while, as is clear in Chapter 6, there was an active policy of decentralizing services and businesses in the Netherlands and elsewhere in Europe in an effort to spread wealth more evenly. Periods of decentralizing were followed by periods of concentrating people and economic activities.

8.2 Myrdal's theory of cumulative causation

In his book *Economic Theory and Underdeveloped Regions* (1957), Swedish economist and Nobel Prize winner Gunnar Myrdal introduced the principle of cumulative causation, i.e. certain regions become the focus of economic activity, thus increasing spatial contrasts (Figure 8.1). Myrdal's assumption was that companies based in wealthy regions have a head start. Wealthy regions have larger markets, allowing businesses to take advantage of economies of scale. They also have other advantages, a better trained workforce and better innovation opportunities. What happens, however, if a region like this drastically expands its output? This initially leads to more local jobs. Assuming there is no unemployment problem, workers from elsewhere are drawn to the new job opportunities, bring their families and swell the local population. The newcomers might well include a relatively large number of highly trained workers. Job migration is a self-selecting process and usually leads to an increase in skilled labour. This alters the business environment, perhaps attracting new economic activities aimed specifically at the skilled workers. This encourages a further concentration of economic activities. The arrival of new businesses can also be motivated by other factors, e.g. the backward linkages of suppliers with the relatively large business that provided the initial growth stimulus. This too has a cumulative effect.

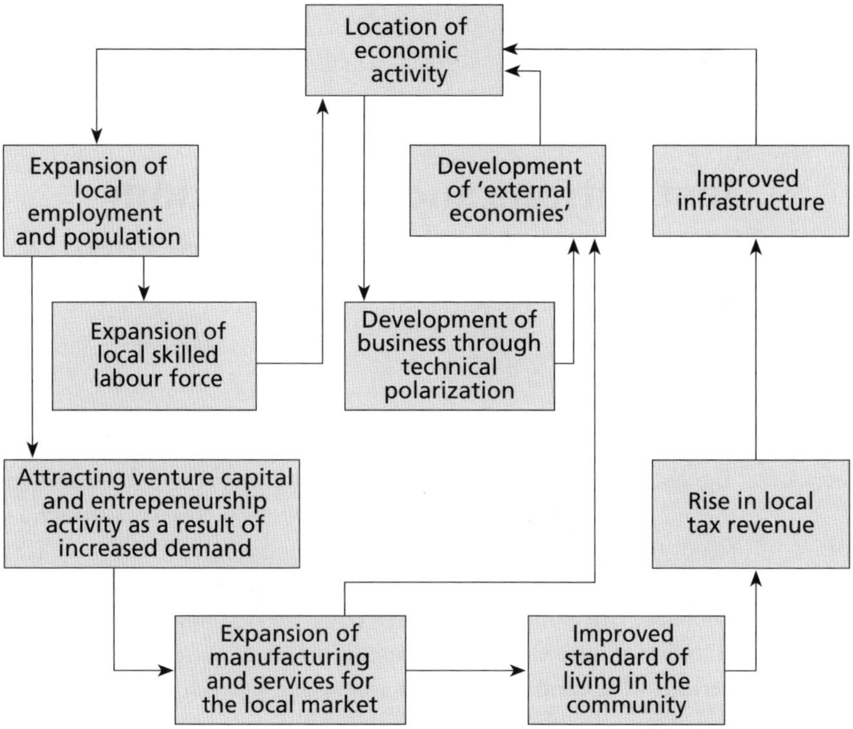

Figure 8.1

The principle of cumulative causation

Box 8.1

Cumulative causation in professional football

Professional soccer is a good example of how cumulative causation works (Dejonghe 2004). Chapter 2 shows how spectator figures can be explained by central place theory. Clubs based in countries and cities with large populations have a competitive advantage. Television rights, income from merchandizing and advertising contracts all depend on the size of the market.

The fact that professional soccer is increasingly market-oriented creates a process of cumulative cycles with more spectators and more income from sponsors. More income from spectators and sponsors allows clubs to pay players higher salaries and attract better players. Better players usually lead to better results. Better results usually mean clubs can charge more for the TV rights and increased spectator figures. In professional soccer, with generally only a handful of successful clubs, long-term success can be seen as an example of Myrdal's process of cumulative causation. The fact that this too is subject to change is illustrated by the rise of regional clubs such as AZ Alkmaar and FC Twente from Enschede in the Netherlands.

A growing local population, with the possibility of an enhanced standard of living as a result of wage rises, constitutes an attractive market for local service providers. Existing providers expand and new ones arrive including bakeries, supermarkets, surgeries, garages, cafés, snack bars etc. As in Christaller's theory, the range of communal facilities such as health care, education and public transport also increases. Their effect on the local business environment can attract other businesses.

Myrdal's theory holds that improvements in a given region's business environment go hand in hand with improvements in its production structure, making it more attractive to businesses and workers. The quality of the production factors and improvements in the business environment generate new economic activity. There are losers as well, since relocating businesses and workers takes something away from the areas they leave. They may lose some of their most valuable businesses and workers and see capital draining away to the core region, where returns are higher. So the growth region's cumulative advantage is accompanied by a cumulative disadvantage for other regions, sharpening the contrast between the core and the periphery. This adverse effect is known as the *backwash effect*. In addition to the cumulative advantages that stimulate prosperity, growth regions undergo another development. As economic development progresses, economic activities fan out, the *spread effect*. It is a result of the rising price of land, since vacant space becomes increasingly rare, the labour market is dogged by labour shortages and rising tension, and congestion and reduced accessibility adversely affect the core region. However, this spatial decentralization does not result in a levelling-off of regional economic variation. There are four reasons why.

- The clustering of economic activity in core regions means labour and space are at an ever higher premium and businesses compete ever more fiercely for them. The weaker companies go under and the remaining ones are forced to be extremely efficient and highly competitive.
- Growth regions can attract new growth stimuli. It is here that the main communication and transport hubs can be found, and their large knowledge and training centres provide them with a wealth of information. The workforce is highly skilled and versatile. All these factors mean changes and innovations are most likely to first manifest themselves in the growth regions.
- In growth regions, economies of scale are maintained by the size of the market. Leaving aside other drawbacks, businesses in peripheral regions are far less able to benefit from economies of scale than their counterparts in core regions. So an entrepreneur will think twice before moving to the periphery.
- Backwash and spread effects can occur on different spatial scales. Backwash effects mainly occur in peripheral regions, where the best workers and most successful businesses drain away. Spread effects occur when growth regions expand. Neighbouring regions benefit since businesses and part of the local labour market from the core region find a new home there.

In other words, even if economic activity spreads out, the original growth area is likely to remain dominant. Its clearest manifestation is a continuous, modern production structure. Myrdal observed an economic divide: strong regions get stronger, weak ones get weaker. The current European economic crisis can be viewed in this light, since notwithstanding regional policies, economic contrasts within the EU are now so sharp a division into Northern and Southern Europe is being considered as a serious option. Yet spatial decentralization can be hazardous to the economic strength of the original growth region, especially in the current economic climate where widespread improvements in communication and transport have weakened the regional links between businesses and the population, certainly in comparison with Myrdal's time.

The theory of cumulative causation is considered valuable in economic geography, but it has also been criticised. Its value lies in its being one of the few dynamic regional development theories. The relation Myrdal saw between the nature of economic activities and regional economic development has been adopted by evolutionary economists (see Chapter 9). Other positive aspects are that Myrdal's emphasized that regional development is anchored in history, and created an overarching development theory linking the behaviour of businesses (e.g. location choice and competitive strategy) with macro-economic structures. So the *theory of cumulative causation* constituted an excellent breeding ground for later studies investigating patterns in regional economic activities within and between nations.

Myrdal's theory also had its share of criticism, which initially focussed on the assumption that once they are weak, regions stay weak, since new growth regions do occur. It should be noted though that once formerly weak regions have transformed themselves into growth regions, the same process of cumulative causation occurs there as well.

A second point of criticism pertains to the link between the behaviour of businesses and interregional differences in economic structures. In the past, how large companies behaved largely confirmed Myrdal's theory, but nowadays it applies less and less. Large companies used to relocate particular facilities, often representing low-value activities, to the periphery. This decentralization did not apply to high-value activities, which often require a high level of interaction, e.g. functions concentrated at a company's headquarters. Nowadays they can also be decentralized, making the pattern ever more complex.

8.3 Perroux's growth poles

Economic base theory and the theory of cumulative causation have led to insights highlighting growth opportunities for regions with weaker economies. The EU and individual nations have policies aimed at boosting the economies of developing regions. This aspect of regional economics was addressed by French economist François Perroux (1950).

According to Perroux, regional economic growth begins with a *firme motrice*, a key company constituting a regional growth pole or *pôle de croissance*. It is a relatively large company in a strong growth industry that maintains intensive linkages with other businesses and industries. In the past, they were mainly industries like car manufacturers, blast furnaces or oil refineries. Nowadays a *firme motrice* can also be for example a university. Via manifold intensive relations with other industries (suppliers, maintenance services, buyers), key companies are able to occupy a central position like a spider in a web. Incidentally, as an economist, Perroux did not focus as much on the regional dimensions of these relations as on the national economy. Governments aiming to stimulate their national economies, he thought, would be wise to invest in key companies. Perroux used the term 'economic space,' as distinguished from geographic, concrete or banal space, terms used by geographers to denote different types of space. It was not until later, mainly in publications by Perroux's followers, that the concept was applied to living geographic space. As of 1960 the term growth pole is mainly used to refer to where the key company is located. This brought Perroux's *growth pole theory* into clear contact with Myrdal's *theory of cumulative causation,* the idea being that a growth pole business in a certain area will lead to overall economic growth there.

In the 1960s, growth pole theory was very popular with regional policy-makers, and many policies to stimulate regional economic growth were based on its principles. It was assumed that a region boasting a growth pole would be able to maintain all its growth effects as a result of four *polarization* or *multiplier mechanisms*:

- *Technical polarization.* This applies to businesses the key company is connected with via buyer/seller relations. The assumption is that some of the other businesses will settle near the key firm.
- *Income polarization.* Increased employment in the key firm and the businesses drawn in via technical polarization generates extra income that new and existing firms benefit from.
- *Psychological polarization.* The establishment of a renowned key firm and subsequent arrival of multifarious businesses create a sense of optimism across the region. Externally, the region's image receives a boost, attracting yet more businesses.
- *Geographic polarization.* The above three types of polarization generate a positive change in the region's business environment, which is beneficial to future growth.

Growth pole theory contains elements already observed in economic base theory, Keynesian multiplier theory and cumulative causation theory, and has been widely criticised, e.g. for barely addressing the location choice of the key firm. Like the theory of cumulative causation, growth pole theory explains what happens when a key firm establishes itself in a region. But to truly answer that question, location theory is called for. It shows that it is not easy to develop a growth pole in an economically weaker and usually peripheral region. Despite large-scale investments (e.g. steel production or chemical plants), many peripheral regions, such as Mezzogiorno in southern Italy, have not witnessed the desired polarization effects or image enhancements and are often labelled 'cathedrals in the desert.' The regional business environment did not have much to offer and many of the polarization effects solely affected other regions, such as the economically stronger north in the case of Italy. The positive effects on the peripheral regions were limited to income polarization and occasionally some psychological polarization.

8.4 Can intervention prevent regional contrasts?

State intervention should always place communal interests above individual ones. The complex task of any government is to assess and balance the various interests and take anticipated developments into account. As the welfare state developed, governments increasingly appropriated this task. Especially in economic circles, some say state interference in society (*big* government) has gone too far. This is a recurring debate, with people in the 21st century increasingly looking to neoliberalism for solutions.

Regional economic policy has undergone a shift towards spatial economic policy and this is a clear illustration of this political reorientation. Instead of a regional economic policy to promote social justice, the Dutch government now has a spatial economic policy with every region making a maximum contribution to the international competitiveness of the nation's industries. In terms of policy objectives, this marks a clear shift from *equity* (equal distribution) to *efficiency*. The main policy instruments have shifted from detailed locally varying subsidies (e.g. investment incentive schemes) to large-scale projects such as the expansion of transport hubs and fast connections with the hinterland.

The same economic geography questions crop up in government policy and spatial economics:

- Can the authorities influence businesses' *choice of location*?
- Can the authorities create a *business environment* that can contribute to regional economic growth?

Even if the answer to both questions is yes, this does not necessarily mean the state would want to intervene. We know how tricky this can be. The choice of location is intimately connected to a wide range of internal factors, some of which are unpredictable and uncontrollable, and is increasingly dependent on decisions made on the international stage, which are similarly difficult to influence. The aims of big government are based on underlying aims and perceptions of the actual problem. For example, the key aim of Dutch regional economic policy in the 1970s and '80s was to reduce interregional variation in living standards.

Clear examples of state intervention in the Netherlands can be found in the ABC policy. Developed as part of the Planning Policy Statement 4 (1988), it is now largely obsolete. The basic premise was that businesses could not simply set up shop wherever they liked but were given A, B or C locations. A locations were main transport hubs such as large train stations, B locations had good access by public transport and automobile, and C locations were situated along motorways. The type of location businesses could choose depended on the mobility needs of their employees/visitors and the accessibility of the location. The overall policy aim was to reduce the use of the automobile to improve living conditions and the environment. In practice the policy was virtually unenforceable and businesses and provincial authorities alike resisted the rigid rules and drastic parking regulations that came with it.

In addition to the planning policies stipulating specific industrial locations for specific businesses, the authorities influenced businesses' choice of location in other ways. The clearest illustrations were the planning schemes for new business parks, the reorganization of older ones and the regional and local traffic and transport policies. Regional and local authorities can make their policies invite businesses rather than put them off so they move elsewhere. One way to do so is by positively influencing the business environment. Local authorities can drastically increase the number of business parks and enhance them with 'extras.' From a game theory perspective (see Chapter 4), this is a logical strategy for individual municipalities but would be disastrous on a national scale, since it would ultimately lead to widespread vacancy.

Box 8.2

Attracting businesses: Do financial incentives work? The case of HP in Friesland in the north of the Netherlands

The history of the Friesland International Business Park (IBF) in Heerenveen in the northwestern Dutch province of Friesland is an example of state intervention and its limitations. Work on this 240-hectare business park was launched in 1993 in an effort to attract large international enterprises. In 1997, Mayor P. de Jonge spoke of 'a dream come true' when the American multinational SCI appeared on the scene. Later that year, SCI chose IBF as the assemblage and distribution hub for Hewlett Packard computers in northwest Europe. In actual fact it was not HP but SCI that carried out the activities, but this did not dampen the spirits. Optimism was the name of the game, and in a matter of years, jobs would increase from 500 to 2000.

However, a mere five years later, SCI left as suddenly as it came, leaving Friesland feeling deflated and millions of euros of subsidy money poorer. The area was once again an ordinary business park. The overall cost of its reorganization, including a nearby industrial zone, was nearly 13 million euros. The closure of this international business park demonstrates the limited influence a government can exercise over the economy and how careful it has to be with financial incentives.
Source: *Reformatorisch Dagblad* 2004

Regional and local authorities try to make their areas attractive places to live and work. They can also be partners in collaborative schemes by businesses, officials and knowledge institutions such as universities to help set up a regional system to boost innovation. The basic idea is to stimulate the growth of new enterprises and help existing ones become more innovative. All this can have a positive influence on the region's image and encourage other businesses to settle there (see Chapter 7). High-quality educational and other facilities such as hospitals, sports and the retail trade can create pull factors for new businesses from other regions and other countries.

A term sometimes used in this context is a country or region's *institutional thickness* (Amin and Thrift 1995), i.e. effective climate of collaboration between a variety of agents in a region such as the business community, institutions and authorities. Several factors determine a region's institutional thickness, e.g. a communal platform consisting of the various regional stakeholders and regular interaction between them, the formulation of a common objective and a broad awareness of the need to work towards a common goal. Depending on the region, the development of institutional thickness can be a laborious process. Stakeholders in wealthy regions often lack an awareness of the need to stimulate the region's institutional thickness.

8.5 State intervention via public policy: The tools

Governments have certain tools at their disposal to intervene in the economy, some direct, some indirect. Subsidies to encourage businesses to implement transport schemes are an example of direct intervention. Indirect intervention is more about exercising structural influence and is aimed at long-term changes in a region's economy, e.g. policies on technology aimed at strengthening the regional economy by introducing innovations. In the past, state involvement in the high-tech defence industry through contracts and subsidies could be considered direct intervention, though most present-day policies deal with imperfections on the knowledge, labour and capital markets to facilitate the implementation of technological innovations.

Most state intervention tools appear to pertain to non-economic aspects of the economy. A few economic stimuli will not be enough to transform a nation of merchants and traders like the Dutch into an industrial or high-tech society. It requires enormous shifts in mentality and focus, such as increasing interest in the sciences. So in addition to direct and indirect intervention tools, we can also draw a distinction between economic and non-economic intervention tools, although the difference between the two is not always clear. Examples of non-economic intervention tools are laws and planning regulations and using them, governments can influence the structure of the economy and the regional economy in particular and the behaviour of economic agents as shown in this scheme.

	Economic	Non-economic
Direct	R&D subsidies	Transport schemes
Indirect	Innovation policies	Location policies

8.6 Market organization and planning

The use of intervention tools leads to market organization. Organizing the market is not the same as disrupting the market. Only in a purely classical, theoretical world do taxes, subsidies, social welfare and regulation constitute disruptions. In reality, markets rarely function properly and need correction. Not every economist will dismiss state intervention as being pointless ideology. In fact it is highly defensible on the basis of economic theoretical motives. Even neoclassical economists assume markets that fulfil certain conditions, e.g. perfect competition and abundant high-quality information, occasionally require some correction by the government. For society in general, this means yields are higher than if the economy has been rigid due to commercial or trade union cartels or monopolies. If the common good is at stake, governments can try to intervene, e.g. via competition policy or by appointing a civil servant to monitor proceedings on the weekly market. So market organization and the free market are certainly not mutually exclusive.

In his 2001 inaugural lecture, Prof. Pieter Tordoir addressed the issue of market organization and the free market and asked, 'What is the relation between the free market's self-regulating invisible hand and the visible hand of planning procedures?' In Tordoir's view, location and development issues pose two fundamental problems for neoclassical economics. Firstly, they always interpret distance as a cost and recommend removing the physical (spatial) and legal barriers, e.g. by constructing transport infrastructure or liberalizing trade. Their perfect world is one free of spatial frictions. But 'space' means more than just distance. It also encompasses the natural world, landscape and environment, all highly valued in modern-day planning. Space represents cultural values worthy of protection, if necessary by state intervention. If this common interest hampers a region's growth, leading to a drop in income, logic dictates that it should be compensated for by a better quality living and working environment and a green and pleasant landscape. Whether this assessment is actually made in government policy is a political decision.

In an economic sense as well, space means more than just distance. Returns to scale and external effects (increasing returns) create a clustering of economic activities by means of agglomeration. Here as well, state intervention comes into play via spatial interventions to facilitate this agglomeration. A good example is all the traffic arrangements needed to maintain access to and from a major airport. Other examples are the ample parking facilities near shopping centres and the knowledge centres or science parks that allow a region to benefit economically from the influx of knowledge. In addition to improving the well-being of the population by safeguarding the environment and creating green spaces, state intervention means optimizing the benefits of certain business locations and the business environment. In his inaugural lecture, Tordoir qualifies the contrasts between the free market and market organization and between economic development and planning. He advocates constructive collaboration, which manifests itself in the discipline of economic geography.

8.7 Conclusion

In a complex society such as ours, more and more activities are carried out by institutions. The institutions' functioning plays an important role in a region's attractiveness to businesses. Countries with weak institutions and plagued by corruption will have difficulty attracting foreign investments.

Whether governments should intervene in the economy has been the subject of many a heated debate. This is understandable, since Myrdal's theory of cumulative causation shows how some regions grow wealthy while others remain peripheral, and the imbalances are difficult to tackle. Growth poles and an active decentralization policy are ways governments have tried with varying degrees of success to reduce these regional imbalances.

9

Evolutionary growth theory

9.1 Introduction

Chapter 8 shows how after a period of cumulative growth, regions can either land in a positive or a negative spiral. Regions with a one-sided production structure are apt to wind up in a negative spiral, the market saturated with the goods produced there. This is expressed in the *product life cycle*. Formerly flourishing industrialized regions like the British Midlands or the Belgian Walloon have been receiving financial aid from the EU for years because they fell behind economically.

The overall economic situation is one of stability with businesses not in the habit of often moving. Paris plays a vital role in the French economy, it always has and almost certainly always will. The same applies to London in England and the Randstad in the Netherlands. Forces seem to be at work in the economic space that ensure the continuation of existing structures. Research into patents and innovation demonstrates an extraordinarily consistent pattern. According to a study by Acs et al. (1994), in 1982 a mere ten U.S. states were responsible for over 80% of the new innovations, and one by Andersson and Koster (2011) demonstrates that in the long-term, some regions generate far more new businesses than others.

9.2 Boom and bust: The theory of long economic cycles

In the evolutionary approach to economics, time is crucial. This leads us to economic waves (see Chapter 5), particularly ones spanning long periods known as *Kondratiev waves*.

Box 9.1

Kondratiev's *long waves theory*, and how it became the death of him

In 1920, after a short stint working for the Russian government at the age of 25, Russian economist Nikolai Kondratiev (1892-1938) founded the Institute of Conjuncture, a centre for the study of business cycles. Born into a peasant family, he was one of the men behind the first Russian agricultural five-year plan. His essay on long economic cycles was published on 6 February 1926 and expanded into a book in 1928. Following extensive research, Kondratiev concluded that economies moved in waves of about fifty years. Each of the long waves (we are currently in the fifth) is associated with a fundamental technical breakthrough. The waves themselves can be divided into seasons. Winter represents a depression, autumn a recession, spring a recovery and summer a period of economic boom. Using data going back to 1789, Kondratiev based his analysis on the prices of raw materials such as iron, and on the wages of agricultural workers in England, France and the U.S.
Kondratiev predicted the Depression of the 1930s, but interestingly enough, his theory also said the capitalist system would survive and recover. This went directly against the communist conviction that the inevitable fate of capitalism was to be overthrown by the proletariat and replaced by a communist utopia. In 1930 he was arrested and sent to a work camp. He was executed on 17 September 1938. However, his theory lives on and still has many proponents.

Figure 9.1

Nikolai Kondratiev (1892-1938)

Source: Torahtimes.com

In 1972, banker David Rosenau and journalist James Schuman introduced Kondratiev's theory into modern economic thinking. Fifty years after the 1929 Wall Street Crash, they predicted the same would happen in 1979. It didn't until 1987, but this did not reduce the renewed interest in Kondratiev's ideas. In 1992, the International N.D. Kondratiev Foundation was founded by the Russian Academy of Sciences to stimulate and reward research in the same vein.

Kondratiev himself did not explain why the economy moves in cycles of growth and contraction, but his later proponents held that the driving force behind the cycles was innovation. The first upward Kondratiev wave was associated with the invention of the steam engine and the onset of the Industrial Revolution. However, as the steam engine was introduced across the board, economies contracted, resulting in a worldwide downward wave.

The second wave was linked to improvements in infrastructure and transport networks, the construction of railways and introduction of steam-powered ocean liners. The third wave was triggered by the invention of the combustion engine and the fourth by the post-World War Two durable synthetic materials. The widespread introduction of semi-conductors and information and communication technology is generally seen as triggering the fifth wave (see Figure 9.2).

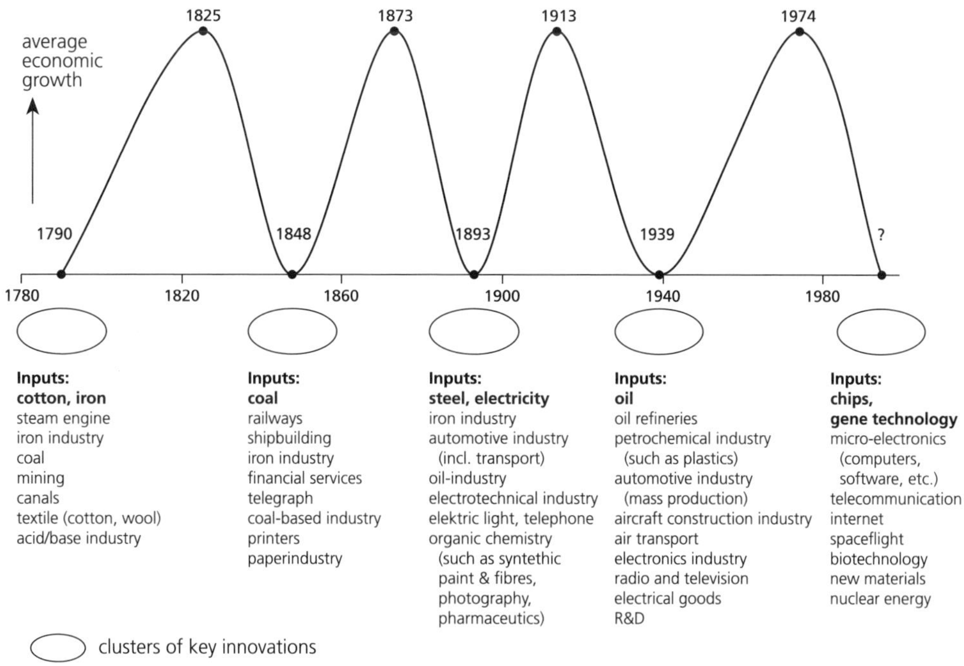

Figure 9.2

Kondratiev's long wave theory

Source: Boschma, Frenken and Lambooy 2002

Innovation played a significant role in Schumpeter's theory. What made it evolutionary is that it was based on the notion that economic growth mainly results from a rise in production caused by the introduction of new, more efficient combinations of factors of production (inputs) and new enterprises and activities. In Schumpeter's approach to innovation, the time factor closely parallels economic waves. Economic recessions, in Schumpeter's view, stimulate and even challenge entrepreneurs to innovate. Innovations can be profitable in these periods because the market is saturated and ready for something new. New products tend to trigger imitations and speed up the ensuing economic growth. Eventually this enables the economy to climb out of its slump and begin a new boom period.

The 2008 and subsequent recessions fit neatly into Kondratiev's system, implying that some day, recovery would come with the advent of a technology that will be fundamentally new but is currently impossible to predict.

Although Schumpeter also addressed other forms of capitalism, for our purposes his approach to *competitive capitalism* is the most interesting. In this variant of capitalism, innovations are introduced in the manner described above by entrepreneurs starting new businesses. According to Schumpeter, a national or regional economy benefits from a political system that fosters the freedom and creativity of entrepreneurs or at least does not hamper them, as in the former planned economies of Eastern Europe.

One more spatial translation of Schumpeter's ideas needs to be made. If conditions for innovation are consistently better in certain regions, they will be able to maintain their favourable position. The assumption in evolutionary economics is that in any given region, not all companies will successfully adapt to the changing business environment. Evolution means selection and only those that adapt will survive. This approach advocates *evolution*, not *revolution*. In the context of this book, the most relevant aspect of evolutionary economics is the idea that regional patterns of economic activity and growth display great stability. Areas that are traditionally part of core regions are likely to maintain that status. This is because they have managed to secure a central position in the national and particularly the international economic world and are the most innovative and entrepreneurial, have the most varied production structure, and benefit most from new transport systems. All of which protects the regional economy from turbulence in the business environment. By influencing the environment, regional authorities can stimulate the process of adaptation. Chapter 5 discusses the evolutionary approach referring to Nelson and Winter (1982) and examines how businesses behave. Core concepts include bounded rationality, the key role played by technological innovation, and the notion of market competition as an evolutionary process of variation and selection.

The success of regions like Silicon Valley and the Third Italy is ascribed to their ability to innovate. This derives from their strong networks of small and medium-sized enterprises (SMEs). A crucial factor is proximity, which fosters the collective learning process. The main reason for this is the workers' high level of mobility. In geographically constrained, rapidly growing and tightly knit entrepreneurial communities, the leap from one company to another is easily made, they can learn from each other and innovation is continuously stimulated.

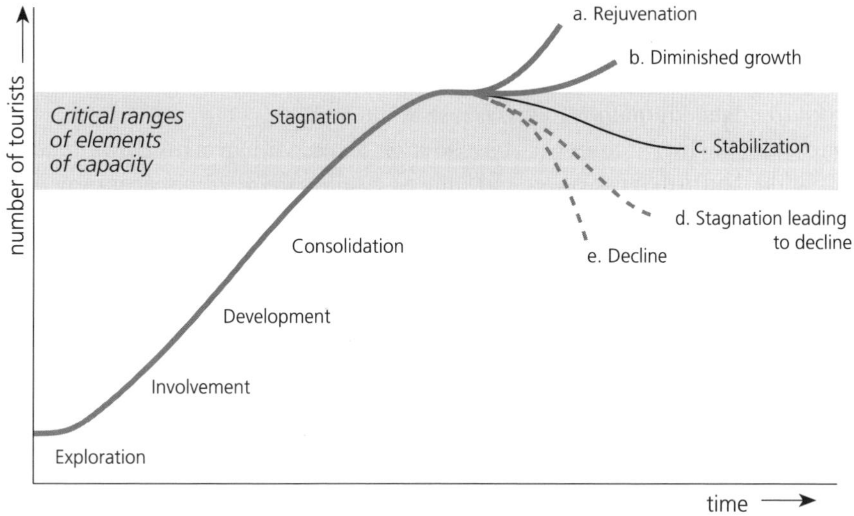

Figure 9.3
The life cycle of a tourist destination
Source: Butler 1980

We have discussed evolutionary models such as the *industry cycle* and *product life cycle* and described the rise and fall of industrial sectors. In 1980, British geographer Richard Butler applied the product life cycle to tourist destinations. The *exploration phase* of his *tourism area life cycle* pertains to relatively unknown destinations with limited facilities visited by only a few adventurous, independent tourists. They use the local facilities and interact frequently with the local population. Daily life there is virtually unaffected by the advent of tourism and the economic and social effects of individual tourists are negligible.

Once tourism figures begin to rise and patterns become discernible, *the involvement phase* begins. Some local residents become involved with tourists and start to provide facilities for them. Interaction between tourists and residents becomes more frequent, and tourists begin to have a social and economic impact on the people involved with them. In this phase, tourism begins to be organized and local authorities may be approached with requests for changes in the local infrastructure to improve tourist access. Bhutan in the Himalayas is currently in this phase.

The *development phase* is next, with regional authorities starting to collaborate with national and international travel organizations to produce targeted place marketing. There is less and less local control over the development of tourism. The region's natural and cultural attractions are actively promoted, new attractions are created and small-scale accommodations are replaced by larger ones, including chain hotels. The look and feel of the region change dramatically and not all the locals are happy with the changes. In the peak season, tourists can outnumber local residents and extra workers need to be brought in. The region also begins to exhibit more variation in terms of types of tourists. Vietnam is in this phase at the moment.

Then comes the *consolidation phase*, with growth slowing down although tourism figures can continue to rise. As the regional economy is now mainly focussed on tourism, the tourist industry tries to find ways to make the tourist season as long as possible and market the destination to new potential tourists. Chains and tour operators hold strategic positions, and traditional facilities make way for amenities for tourists. The level of tourism may lead to irritation among the more traditional local residents, particularly those with no involvement in it and whose livelihood does not depend on it. The Moroccan city of Marrakech is currently in this phase.

When tourism figures peak, a region enters the *stagnation phase*. It is operating at full carrying capacity, and social, economic and environmental problems begin to emerge. The area has a well-defined image but is no longer fashionable. It relies on repeat visits and specific interest groups such as conferences. There is surplus accommodation capacity and the region's natural and cultural attractions are overshadowed by newly created attractions such as casinos and the hospitality industry. The region's image shifts from popular tourist destination to mass tourism destination. The Spanish Costa Brava and the Turkish Aegean coast are good examples of regions in the stagnation phase.

Once it reaches the *decline phase*, a region is no longer able to compete with newer attractions or other regions, e.g. Turkey's rise as a tourist destination was at the expense of Greece. Tourism figures drop and rather than vacationing, tourists come for day trips and short-term stays. The level of facilities declines, property frequently changes ownership and hotel chains pull out. This happened in coastal resorts like Brighton and Blackpool in the U.K. and Noordwijk in the Netherlands. In his 2011 Master's thesis, Guido Wiemer showed the Dutch town of Ede in the stagnation phase, although he rightly observed that labelling the various phases can sometimes be problematic.

A region can also *rejuvenate* itself. Working together, local entrepreneurs and authorities can sometimes halt the downwards spiral and turn it upwards.

This is what happened at European health resorts like Vichy in France and Karlovy Vary in the Czech Republic. Rejuvenation can occur if all the stakeholders collaborate towards the same end.

In steadily declining areas such as the Belgian Walloon region, the evolutionary approach has something to offer. An analysis of the situation there reminds us of the example of Kodak in Chapter 5. Kodak failed to keep a watchful eye on market developments and change its strategy in time. As a result, it is now headed for bankruptcy while archrival Fuji is flourishing thanks to its timely switch to cosmetics and medication. There is a similar situation in regions with traditional primary industries where the extreme focus on the predominant industry hampers change. This is a *regional lock-in*, within which Grabher (1993) distinguished a *functional lock-in*. In Detroit, the steel industry and automobile industry are so inextricably linked that the relations have become rigid, each leaving the other very little room for change. As suppliers do not need to worry about selling their products, they do not feel a need for market research or marketing. The lock-in is also cognitive in the sense that rather than welcoming external knowledge, new information is made to fit in with the existing internal knowledge (comparable to the cognitive dissonance referred to in Chapter 3). There can also be a *political lock-in*, a kind of pact between regional authorities, employers and workers, who all do everything in their power to maintain the status quo, thus hampering economic renewal.

Figure 9.4
'The Tragedy of Detroit': A special report in the October 2011 issue of Time
Source: *Time*, October 2011

9.3 Clustering as a result of evolutionary processes

It is hard to predict where innovative business activity will occur and chance certainly plays a role. This said, innovation tends to be concentrated in urban regions. Large cities are particularly attractive to services like banks, fashion designers and publishers, and more basic industrial activity can generally be found in smaller towns and cities (Boschma and Frenken 2011).

Box 9.2

Clusters and networks

The issue of why certain activities are spatially concentrated was at the heart of economic geographic research (e.g. Marshall 1920) and later readdressed by Porter (1990). In addition to what is noted in the previous chapters about clusters and networks, there are a few more comments to be made. Clusters are spatial concentrations of related businesses that may or may not have a collaborative interaction. Classic examples are Silicon Valley and the less well-known Third Italy. Clusters emerge as a result of external conditions, natural resources such as locally available raw materials, anthropogenic resources such as airports, or urban resources such as the density, quality and variety of other producers or providers and consumers. Concentration is a condition for clustering but concentration alone is not enough to create genuine clustering. We only refer to clustering if there is a self-reinforcing process among the related businesses. Brenner (2004) distinguished three such self-reinforcing processes:

- direct contact and collaboration between firms;
- the creation of favourable local conditions more and more businesses benefit from as conditions grow more and more favourable;
- the fostering of support services that challenge businesses and help generate better and more innovative products.

Unlike clusters, networks do not require spatial concentration. The term networks mainly refers to businesses that work together. Not every business, incidentally, is interested in being part of a network. Some prefer to operate strategically or financially on their own and are called 'stand-alones.' Others swear by regional informal knowledge exchange ('local buzz'), and highly specialist businesses are often forced to network at the international level ('global pipeline'). It is generally accepted that a combination of local buzz and global pipeline is the most fruitful. Information obtained nearby is cheap and fairly reliable, but an international outlook prevents a business from becoming inward-looking or in evolutionary terms, from lock-in.

Box 9.3
The Third Italy, success through clustering

Italy has a rather negative reputation as a country plagued by soaring national deficits, corrupt politicians, sluggish state companies and an unstable government. Not a good foundation for a flourishing economy, it would seem. Yet despite the sometimes difficult conditions, Italy witnessed strong economic growth in the 1990s, mainly thanks to industrial centres with small and medium-sized family-run businesses in the provinces of Emilia-Romagna, Veneto, Friuli and Tuscany in north and central Italy.

The late 1970s were the years of the Third Italy, following the First Italy (the industrialized north with towns like Milan and Turin) and the Second Italy (the agricultural south). Rather than a geographic area, the Third Italy is characterized by an organizational structure for entrepreneurial activity with small and medium-sized family-run businesses that are successful on the export markets. They tend to be manufacturers of shoes, clothing, furniture, machinery and ceramics.

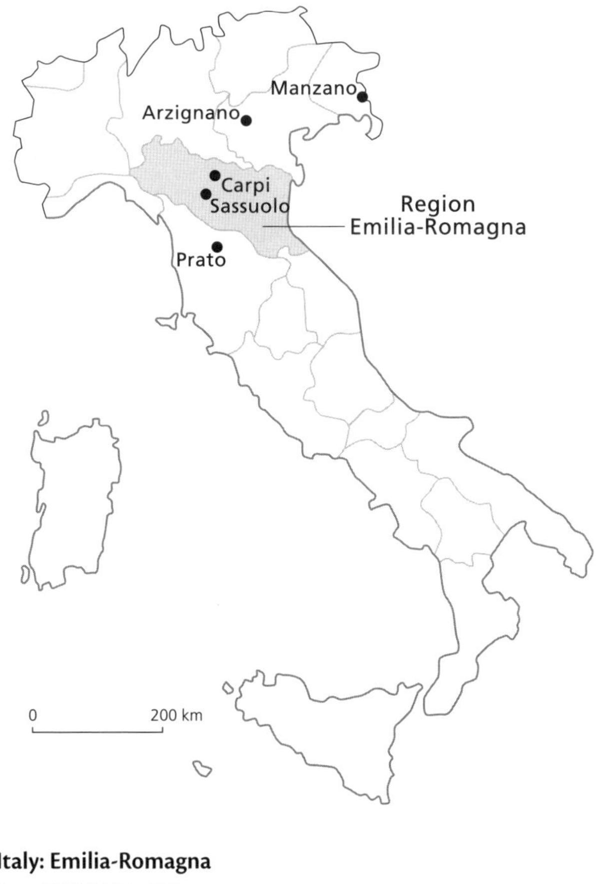

Figure 9.5
The Third Italy: Emilia-Romagna
Source: Williams 1995; Unido 2012

Third Italy businesses can adapt rapidly to market changes and can compete and collaborate with each other. They tend to cluster and the towns of Carpi and Prato are known for their ready-made clothing, Arzignano is famous for its shoes, Manzano is a major centre for furniture production and Sassuolo is home to over 300 manufacturers of ceramic tiles. As a result of family ties and social relationships within the businesses, the workers feel secure in their jobs. They value them more than membership in the often polarizing trade unions that represent the interests of workers in the industrial north. Rather than aiming to achieve economies of scale, First Italy businesses tend to maintain flexible outsourcing. Economic developments in the Third Italy underpin the concept of industrial districts in economic geography as well as Michael Porter's 'diamond.' According to Unido (2012), the Third Italy's success results from:

- geographic proximity of small and medium-sized enterprises;
- specialization per sector;
- predominance of small and medium-sized enterprises;
- high level of collaboration between businesses;
- competition based on innovation;
- a socio-cultural identity based on mutual trust;
- active trade organizations;
- supportive regional and local authorities.

To explain the observed clustering, evolutionary economic geographers analyse the extent to which new businesses emerge and existing ones disappear in the long term. Heebels and Boschma (2011) used a dataset on publishers covering the period from 1880 to 2008. One important outcome was that the Amsterdam publishing cluster had less to do with the city itself than with a number of highly successful publishers that in turn produced successful spin-offs. Once again it is clear that the number of start-ups in a given region greatly depends on the number of existing businesses. This also explains why clusters, once established, have such longevity. Even in the absence of locational advantages, the presence of numerous potential parent companies for spin-offs guarantees continued growth. The trickiest but most interesting question is of course, Why do clusters emerge in one region but not in another? This question is addressed in various ways in every chapter of this book. Classicists and neoclassicists hold that clusters emerge where costs are lowest and yields are highest. Behaviourists hold that image plays a major role. The previous chapter shows how clusters are underpinned by institutions. Evolutionary economists tend to view clusters from the perspective of Klepper's life cycle concept. In his seminal study of the automobile industry, Klepper (2007) concluded that clusters are the result of a snowball process. By inheriting routines that work well, companies that have spun off from a successful parent company stand a better chance of survival than their competitors, and since they tend to settle near the parent company, they reinforce clustering. There are two important consequences of this approach. Firstly, there is regional path dependence at play. This means regions have varying chances of generating a successful new sector. Regions with an abundance of innovative businesses are more likely to undergo continued growth. In short, the local presence of numerous technologically similar industries is a major advantage. The Dutch twin towns of Bunschoten / Spakenburg are a good example, where fishmonger Arie van de Groep turns fish waste and confectioneries past their sell-by date into gas by means of anaerobic digestion, an innovative way of creating renewable energy by fermenting food waste. The fishmonger's anaerobic digester is the largest in the country. With a capacity of 800 cubic metres of green gas, he is able to provide renewable energy for much of Bunschoten-Spakenburg and nearby Eemnes and Baarn. Stedin is the energy company responsible for transporting the gas, and the provincial authorities are investigating possibilities for a biogas hub to connect the various providers of biogas to the natural-gas network. In addition, regional authorities are looking into whether roadside grass clippings can be turned into gas. All of which demonstrates how related sectors can set in motion a whole chain of developments.

Although a variety of related sectors is advantageous, it by no means guarantees success. In addition to path dependence, something else to be aware of is that clusters can also thrive without any discernible locational advantages. Studies into e.g. fashion, sofas, automobiles and automobile tyres have shown that clusters alone cannot enhance a new enterprise's chances of survival. What clinches the matter is mainly whether the spin-off has a successful parent company.

The evolutionary approach sets great store by innovation and research from this perspective focuses mainly on firms with the potential to develop into successful parent companies. Numerous ideas have consequently been developed in this connection. One of them is the incubation hypothesis, first formulated by Hoover and Vernon in 1959 and then elaborated upon by Leone and Struyk (1976). According to these theorists, innovative activity occurs in low-threshold markets that do not require high capital expenditures. To be successful, innovative enterprises need to be able to communicate easily with clients (face-to-face contact) and suppliers, and require specific capacities from their staff, such as a flexible and communicative attitude. Innovative enterprises also benefit from urban locations, where property, labour and transport costs are lower and risks can be kept to a minimum by enabling businesses to settle near their customers and suppliers.

Urban areas in particular are viewed as offering reliable and favourable conditions and in incubation theory, they are considered the ideal setting for innovation. Incidentally, this does not include city centres, where property prices are too steep. For innovative businesses, the combination of information density and low property prices makes older buildings in the zones just outside the centre the ideal location. As Jane Jacobs (1961, p. 188) famously put it, 'Old ideas can sometimes use new buildings. New ideas must use old buildings.' Incubation theory is dynamic and after a while, it says, successful innovative businesses tend to move out of urban centres. The initial locational benefits cease to be advantageous, as businesses make a name for themselves on the market, build up a reputation, develop their own information networks and no longer depend on local information. In addition, their internal growth means they physically outgrow their location on expanding markets operating beyond the local scale, though they still need to safeguard their accessibility to customers and suppliers. Most of them move to the urban periphery or even further out to easily accessible locations, freeing up space for new economic activity. Thus the incubation nature of the original urban location is maintained.

Although nowadays many towns and cities have an incubation policy, incubation theory has been sharply criticised in academic circles (e.g. de Jong 1987). It is often unclear for example what the geographic range is of incubation theory. Are we talking about buildings or blocks, about streets as Jane Jacobs said and Aart Jan van Duren (1995) investigated, or about municipalities or even entire urban regions? Where should the line be drawn? In addition to theoretical and geographical points of criticism, practical objections can be raised. As a result of the urban renewal of post-industrial towns and cities, old factory buildings have either disappeared or been renovated as heritage monuments, driving up the rent. Without hefty state subsidies, they would be unsuitable as incubator spaces. Nevertheless, incubation theory is still relevant and is actively used in the development of science and technology parks on the urban periphery, where we find dedicated incubation facilities.

According to the related seedbed theory, entrepreneurs tend to stay local when they set up new businesses. They are familiar with the local situation and the communities are already familiar with them, facilitating contact with suppliers and customers and making it easier to secure the necessary capital and attract potential staff. Entrepreneurs are apt to set up shop in their home towns, sometimes even in their own homes. Seedbed theory states that the distribution of new economic activity is more dependent on where the entrepreneur lives than on the supply of affordable business property.

In her book *Cities and the Wealth of Nations* (1984), Jane Jacobs stressed the importance of *urbanization economies*, i.e. benefits for companies located in an urban environment. Unlike locational benefits, they pertain to dissimilar rather than similar companies and a different set of urban amenities. As businesses form clusters, the benefits increase. Aart Jan van Duren (1995) used this idea in his PhD research on the development of a number of shopping streets in central Amsterdam. He argued that many specialist retailers benefit from a location near the city centre, allowing them to bask in its glow. The historical centre attracts tourists, which generates footfall for local specialist retailers.

Contrary to Christaller's theory (see Chapter 2), these specialist functions do not end up at the top of the urban centre's hierarchy.

Following Gleaser *et al.* (1992), in his 2002 thesis Frank van Oort contrasted localization (inter-sector knowledge exchange between businesses) with urbanization (intra-sector knowledge exchange between businesses). His research confirms Jane Jacobs' ideas in the sense that industrial sectors located in urban environments with a diverse economy exhibit the greatest employment growth. Incidentally, van Oort noted that in the Netherlands, agglomeration effects occur on a relatively large scale. Urbanization benefits are felt all across most of the Randstad region, or as it is known nowadays, the Delta Metropolis.

Jacobs stated that only cities with decades or even centuries of economic growth can expect to continue to grow in the future. However, she did not confine herself to economic activity, arguing that cultural activity is also of great importance to the economic vibrancy of cities. In this she agreed with Richard Florida (see Chapter 7). The creative industries for example have an important role to fulfil in industrial design. Generally speaking, the wealth of ideas generated in cultural and commercial settings can contribute to a highly differentiated economic, spatial and cultural fabric, which in turn can provide a stimulating environment for a wide range of economic activities.

9.4 Conclusion

In the evolutionary approach to regional economic development, history matters. Whether a region grows economically is more a matter of the business community's ability to adapt to changing circumstances than the region's business environment. Economic growth is cyclical, and the economically successful regions are the ones where businesses are able to adapt to and anticipate the cycles. Regions with multifarious related industries stand a particularly good chance of surviving the changes accompanying economic cycles in terms of innovation and competition. Above all, evolutionary economics teaches us that the future of a region is partly determined by its point of departure, since continuity plays a major role. Successful clusters largely derive from innovative companies that in turn yield successful spin-offs. It is not surprising that local authorities do their utmost to create gems of this kind by providing incubator and start-up facilities.

10

Epilogue

In the previous chapters we address theories and approaches related to three central questions:

- Why does business X prefer location A to location B?
- Why does region A perform better economically than region B?
- To what extent can the state influence business X's location choice and region B's economic development?

We outline multifarious concepts, theories and approaches to give an idea of the possible answers to the questions. The wide array of theories shows there is no such thing in economic geography as a perfect theory or best approach. The theory used depends on what is being studied. Moreover, a research topic can be investigated from any number of theoretical perspectives.

10.1 Economics and geography

The theories and approaches all pertain to the area where economics and geography meet, the border zone sometimes a no-man's land in the history of both disciplines. This is aptly illustrated by the post-World War Two decline in economic geography at a university as prestigious as Harvard. It is all the more striking that it was in Harvard circles that starting in the mid-1980s, authoritative economists and historians began to put economic geography back at the heart of the academic debate. In this context we have introduced historian David Landes and economists Michael Porter and Ricardo Hausman. This revival is surprising since the rise of the Internet led many to predict the demise of economic geography. The age of discovery seemed to be over and physical distance irrelevant. But this illusion that anything can happen anywhere and regional differentiation will disappear turned out to be based on a misconception.

Figure 10.1
Michael Porter

At the heart of the misconception are human beings and the natural world. Natural resources are unevenly distributed, and events like the failure of enormous projects in the former Soviet Union (e.g. redirecting rivers in Siberia and the partial drying up of Lake Aral) demonstrate how unwise it is to tamper with nature for the sake of the economy. The relation between nature and economics is tricky and often determined by culture. Some civilizations flourish in a certain environment, others perform less well under the same conditions, affecting economic growth and living standards. Flourishing economies attract more economic activities, generating a self-reinforcing process. Centres of economic activity become economic powerhouses. The best known example is of course Silicon Valley. This iconic birthplace of information and communication technology actually originated in nearby San Francisco and its explosive growth is the result of fierce competition between a great many ICT companies, all located on the same small, confined patch of land. Here again, geography played a role in economic growth.

10.2 Location choice and regional economic growth

There is a clear distinction between location choice and regional economic growth. Location choice has to do with how a business perceives a region's business environment. The relevant aspect of the business environment consists of factors that differ from region to region. This is why we speak of locational factors. By contrast, regional economic growth has to do with much more than just locational factors. So answering the second question is much harder than explaining location choice. This is probably why in economic geography, theories on location choice are more developed than those on regional economic growth. Although their underlying presuppositions have been superseded, the classical normative location theories (Von Thünen, Weber and Christaller) are still relevant today. We illustrate this using examples from professional soccer, supermarkets and the cardboard industry. Although transport costs are far less dominant now than in Weber's day, they still play a significant role in industries dealing with weight-losing raw materials or ubiquities. This is abundantly clear from the example of the cardboard box industry, where location patterns are still determined by principles formulated by Weber. Christaller's ideas can similarly still be discerned in the fortunes of professional soccer clubs and the threat of hospital closures.

As Lösch knew, there is no such thing as a one-size-fits-all optimal location. The ideal location for a fully automated automobile factory is not the same as for a factory where automobiles are still largely assembled manually. As large businesses generally have a different set of input factors than smaller ones, the optimal location for a major brewery is not the same as for a microbrewery. Additions and refinements often derive from neoclassical approaches to location theory and behaviouralism also adds its perspectives to the mix.

It should be emphasized that despite imitative behaviour, there are considerable differences between different businesses' strategic policies. Many large companies also regularly adjust their strategies, sometimes forced to by the circumstances. This means the same location can represent different values for different companies. Some wouldn't even consider investing in Russia right now, others feel they cannot afford not to. But it also means businesses can come to view a location in a different light. In the 1960s and '70s, Philips could be found in locations all across the Netherlands, most of which have since been abandoned.

Given all these theoretical additions and refinements, it is clear that it is impossible to formulate one grand location theory. Every choice is unique. The elements contributing to this unique choice are described in the theories and approaches discussed in this book.

10.3 Regional growth and development

Three central aspects can be distinguished in regional growth theories:

- growth and regional specialization;
- regional convergence or divergence;
- regional collaboration.

All the approaches implicitly or explicitly assume that to achieve growth, regions have to take advantage of their strengths. This implies building on the activities that represent a region's strengths and reinforcing regional specialization. Ricardo's comparative advantage theory and Porter's model are clear examples of this type of approach. It is also evident, albeit implicitly, in growth pole theory. In this approach, key businesses create cumulative economic growth in a certain region. Clearly, boosting the economy of a peripheral rural region and a region such as Greater Amsterdam requires very different types of key businesses. To play this role, businesses have to fit in with a region's possibilities, its strengths.

After an initial boom, over-specialization can however create problems. Vernon's interpretation of the product life cycle is relevant in this context. A good example is the textile industry in the eastern Dutch region of Twente. Not only can a fall in demand affect a regional economy. Over-specialization can lead a region towards economic and institutional lock-in, resulting in a lack of innovation. This is evident in Nelson and Winter's evolutionary approach, and Jacobs' discussion of cities as centres of innovation is similarly relevant in this context. A prerequisite for innovation is a highly diversified economy, and many regions with a traditionally vibrant economy are still going strong. Nelson and Winter's evolutionary approach explains this from a historical perspective as well as in terms of path dependence. In trying to determine whether you are dealing with convergence or divergence, it is important to remain aware of the scale. Relations between countries in central Africa are played out on a totally different scale than between urban and rural regions in the Netherlands.

Economic geography has always had a strong practical focus in the Netherlands, and this book testifies to this. The role of the state in economic planning has changed. In their regional economic policies, the authorities always try to balance meeting the aims of social justice (raising living standards across the board) with being economically effective (concentrating on achievement). Since the 1980s, the Dutch pendulum has swung towards economic effectiveness. Investments in the seaports, connections to the hinterland and the main Randstad cities are key examples of this policy. At the same time, following European-level regional policy, the Dutch authorities stimulate rural economies like those in the north or in the province of Flevoland. For a variety of reasons, there is often a mismatch between the market and the authorities. Nowadays, the branch of economic geography concerned with policy development focuses on creating concepts to reconcile private economic market forces and collective spatial organization. In this respect as well, economic geographers build bridges.

Literature

Aalst, I. van, Atzema, O., Boschma, R., Heinz, F. & Oort, F. van (2005). *Creatieve klasse en regionaal-economische groei.* Den Haag: Ministeries van Onderwijs, Cultuur en Welzijn en Economische Zaken.

Acs, Z.J., Audretsch, D.B. & Feldman, M.P. (1994). 'R&D spillovers and recipient firm size'. *Review of Economics and Statistics,* vol. 76, no. 2, pp. 336-340.

Akerlof, G. (1970). 'The Market for "Lemons". Quality Uncertainty and the Market Mechanism'. *Quarterly Journal of Economics,* vol. 84, no. 3, pp. 488-500.

Alonso, W. (1965). *Location and land use. Toward a general theory of land rent.* Cambridge: Harvard University Press.

Amin, A. & Thrift, N. (1995) 'Globalisation. Institutional thickness and the local economy', *Managing Cities. The New Urban Context.* Eds. P. Healey, S. Cameron, S. Davoudi et al., John Wiley & Sons, Chichester, pp. 91-108.

Andersson, M. & Koster, S. (2011). 'Sources of persistence in regional start-up rates. Evidence from Sweden'. *Journal of Economic Geography,* vol. 11, pp. 179-201.

Annoni, P. & Kozovska, K. (2011). *EU regional competitiveness Index 2010.* Brussel: Europese Commissie, Joint Research Centre. [consulted] 30 januari 2012 [through] http://composite-indicators.jrc.ec.europa.eu/Document/RCI_EUR_Report_updated.pdf

Ashworth, G. & Kavaratzis, M. (2009) 'Beyond the logo. Brand management for cities'. *Journal of Brand Management,* vol. 16, no. 8, pp. 520-531.

Ashworth, G. & Page, S.J. (2011) 'Urban tourism research. Recent progress and current paradoxes'. *Tourism Management,* vol. 32, no. 1, pp. 1-15.

Ashworth, G.J. & Voogd, H. (1988). 'Marketing the city. Concepts, processes and Dutch applications'. *Town Planning Review,* vol. 59, no. 1, pp. 65-79.

Balthelt, H. & Glucker, J. (2002). *Wirtschaftsgeographie.* Stuttgart: Ulmer.

Barschot, J. van (2003), 'Een doos mag nooit duur zijn'. *NRC Handelsblad,* 18 februari 2003, p. 16.

Berry, B.J.L. & Chauncy, D.H. (1970). 'Walter Christaller. An Appreciation'. *Geographical Review,* vol. 60, no. 1, pp. 116-119.

Boisen, M. (2008) 'Cities are not products, so stop trying to sell them as such'. *International congress on Place Branding,* Berlijn, 4-6 december 2008.

Boschma, R.A. (1997) 'New industries and windows of locational opportunity. A long-term analysis of Belgium'. *Erdkunde,* vol. 51, no. 1, pp. 12-22.

Boschma, R.A. (2005) 'Role of proximity in interaction and performance. Conceptual and empirical challenges'. *Regional Studies,* vol. 39, no. 1, pp. 41-45.

Boschma, R. & Frenken, K. (2011) 'The emerging empirics of evolutionary economic geography'. *Journal of Economic Geography,* vol. 11, no. 2, pp. 295-307.

Boschma, R.A., Frenken, K. & Lambooy, J. (2002) *Evolutionaire economie. Een inleiding.* Bussum: Coutinho.

Brakman, S., Garretsen, H. & Marrewijk, C. van (2006) 'Comparative advantage. Cross-border mergers and merger waves. International economics meets industrial organization'. *CESifo Forum,* vol. 7, no. 1, pp. 22-26.

Brakman, S., Garretsen, H. & Marrewijk, C. van (2009) *An Introduction to Geographical Economics.* Cambridge: Cambridge University Press.

Brenke, K. & Wagner, G. (2007) 'Zum volkswirtschaftlichen Wert der Fussball-Weltmeisterschaft 2006 in Deutschland'. *Research Note 19,* Deutsches Institut für Wirtschaftsforschung, april 2007, Berlijn. [consulted] 12 maart 2012 [through] http://www.diw.de/documents/publikationen/73/diw_01.c.56559.de/rn19.pdf.

Brenner, T. (2004) *Local Industrial Clusters. Existence, emergence and evolution.* Londen/NewYork: Routledge.

Burgess, J.A. (1982). 'Selling places. Environmental images for the executive'. *Regional Studies,* vol. 16, no. 1, pp. 1-17.

Butler, R.W. (1980) 'The concept of a tourist area cycle of evolution. Implications for management of resources'. *Canadian Geographer,* vol. 24, no. 1, pp. 5-12.

Buursink, J. (1971). *Centraliteit en hiërarchie. De theorie der centrale plaatsen in enkele Nederlandse industriegebieden.* Assen: Van Gorcum.

Castells, M. (2000). *The Rise of the Network Society.* The Information Age: Economy, Society and Culture, vol. 1 (second edition, first edition 1996). Cambridge/Oxford: Blackwell Publishers.

Christaller, W. (1933). *Die zentralen Orte in Süddeutschland. Eine ökonomisch-geographische Untersuchung über die Gesetzmässigkeit der Verbreitung und Entwicklung der Siedlungen mit städtischen Funktionen.* Jena: Gustav Fischer.

Churchill Semple, E. (1911). *Influences of Geographic Environment.* Charleston: BiblioBazaar.

Crevoisier, O. (1999). 'Two ways to look at learning regions in the context of globalization. The homogenizing and particularizing approaches'. *Geojournal,* vol. 49, no. 4, 1999, pp. 353-361.

Cushman and Wakefield Global real estate solutions (2011). *European Cities Monitor 2011.* [consulted] 20 januari 2012 [through] http://www.cushwake.com/cwglobal/jsp/kcReportDetail.jsp

Dalen, H.P. van & Vuuren, A.P. van (2003). 'Hoe groot is de transactiesector?'. *Economisch Statistische Berichten,* vol. 88, no. 4412, pp. D9-D10.

Damme, E. van (2003). 'Intermediair: mens of markt'. *Economisch Statistische Berichten,* vol. 88, no. 4412, pp. D13-D15.

Dejonghe, T. (2004). 'Restructuring the Belgian professional football league. A location-allocation solution'. *Tijdschrift voor Economische en Sociale Geografie,* vol. 95, no. 1, pp. 73-88.

Dejonghe, T., Hoof, S. van & Kemmeren, T. (2006). *Voetballen in de kleine ruimte. Een onderzoek naar de geografische marktgebieden en ruimtelijke uitbreidingsmogelijkheden voor de clubs in het Nederlandse betaald voetbal.* Nieuwegein: Arko Sportmedia.

Dicken, P. (2011). *Global Shift. Mapping the contours of the world economy.* London: Sage.

Driessche, W. van (2010). 'Het yin en yang van de vrije markt'. *De Tijd,* 13 februari 2010, p. 39. [consulted] 5 april 2012 [through] http://www.academic.lexisnexis.nl/.

Dunning, J.H. (1988). *Explaining international production.* London: Unwin Hyman.

Duren, A.J. van (1995). *De dynamiek van het constante. Over de flexibiliteit van de Amsterdamse binnenstad als economische plaats.* Utrecht: Universiteit Utrecht.

Economist, The (May 31, 2001) "Malcolm Mclean". *Economist.* [consulted] 12 november 2011 [through] http://www.economist.com/node/638561.

Ernst & Young (2011) *Barometer Nederlands Vestigingsklimaat.* [consulted] 12 Oktober 2011 [through] http://www.ey.com/Publication/vwLUAssets/Barometer_Nederlands_vestigingsklimaat_2011/$FILE/Barometer%20Nederlands%20vestigingsklimaat%202011.pdf

Feddersen, A. & Maennig, W. (2010). *'Sectoral Labour Market Effects of the 2006 FIFA World Cup'.* Working Papers 033, Chair for Economic Policy, Hamburg: University of Hamburg.

Florida, R. (2002). *The rise of the creative class.* New York: Basic Books.

Friedmann, T. (2005). *The world is flat. A brief history of the twenty-first century.* New York: Farrar, Straus and Giroux.

Fukuyama, F. (2011). *The origins of political order.* Londen: Profile Books.

Gertner, D. (2011). 'A (tentative) meta-analysis of the place marketing and place branding literature'. *Journal of Brand Management,* vol. 19, no. 2, pp. 112-131.

Ghemawat, P. (2001). 'Distance Still Matters. The Hard Reality of Global Expansion'. *Harvard Business Review,* vol. 79, no. 8, pp. 137-147.

Ghemawat, P. (2011). *World 3.0. Global Prosperity and How to Achieve It.* Boston: Harvard Business Review Press.

Glaeser, E. (2011). *Triumph of the City.* Londen: The Free Press.

Grabher, G. (1993). *The embedded firm. On the socioeconomics of industrial networks.* New York: Routledge.

Groenewegen, J. (2012). 'Fuji keek vooruit en won dus wel'. *NRC Handelsblad,* 27 januari 2012, [consulted] 31 januari 2012 [through] http://academic.lexisnexis.nl/.

Hakanson, L. (1979). 'Towards a theory of location and corporate growth'. *Spatial analysis, industry and industrial environment* (eds. F. Hamilton & G. Linge), pp. 115-138. Chichester: Wiley.

Hausmann, R. (2001). 'Prisoners of Geography'. *Foreign Policy,* no. 122, January-February 2001, pp. 44-53.

Heebels, B. & Boschma, R. (2011) 'Performing in Dutch book publishing 1880-2008. The importance of entrepreneurial experience and the Amsterdam cluster'. *Journal of Economic Geography,* vol. 11, no. 6, pp. 1007-1029.

Heijman, W.J.M. (2002). *Regionale economie. Van vestigingsplaatstheorie naar regionale ontwikkeling.* Delft: Uitgeverij Eburon.

Hofstede, G. (1980), *Cultures consequences.* Newburry Park: Sage.

Hofstede, G. (1991). *Cultures and Organizations. Software of the Mind.* London: McGraw Hill.

Hofstede, G., Hofstede, G.J. & Minkov, M. (2011). *Allemaal Andersdenkenden. Omgaan met cultuurverschillen.* Amsterdam: Atlas/Contact.

Hoof, S. van (2012). 'Meten om te weten'. *Geografie,* jrg. 21, no. 1, pp. 32-35.

Hoover, E. & Vernon, R. (1959). *Anatomy of a metropolis. The changing distribution of people and jobs within the New York Metropolitan Region.* Cambridge : Harvard University Press.

Hospers, G.J. (2009). *Een kleine geografie van het geluk.* inaugural lecture as professor citymarketing at Radboud Universiteit Nijmegen. Nijmegen: Radboud Universiteit.

Hotelling, H. (1929). 'Stability in competition'. *The Economic Journal,* vol. 39, pp. 52-79.

Hottes, R. (1983). 'Walter Christaller'. *Annals of the Association of American Geographers,* vol. 73, no. 1, pp. 51-54.

Huntington, E. (1924). *Civilization And Climate.* Yale: Yale University Press.

IMD Institute for Management (2011). *The World Competitiveness Yearbook 2010.* Lausanne: IMD Institute for Management.

Jacobs, D., Boekholt, P. & Zegveld, W. (1990). *De economische kracht van Nederland.,* Den Haag: SMO.

Jacobs, J. (1961). *The dead and life of great American cities.* New York: Random House.

Jacobs, J. (1984). *Cities and the Wealth of Nations. Principles of economic Life. New York:* Vintage.

Janssen, R., (2011). 'Blunderen ten tijde van crisis'. *NRC Handelsblad,* 7 oktober 2011. [consulted] 5 april 2012 [through] http://www.academic.lexisnexis.nl/.

Jong, M.W. de (1987). *New economic activities and regional dynamics* (dissertatie). Amsterdam: Universiteit van Amsterdam.

Kahneman, D. & Tversky, A. (1979). 'Prospect Theory. An Analysis of Decision under Risk'. *Econometrica,* vol. 47, no. 2, pp. 263-291.

Kavaratzis, M. & Ashworth, G.J. (2005). 'City branding. An effective assertion of identity or a transitory marketing trick?'. *Tijdschrift voor Economische en Sociale Geografie,* vol. 96, no. 5, pp. 506-514.

Keuning, H.J. (1969). *De denkwijze van de sociaal-geograaf.* Utrecht: Aula.

Klepper, S. (2007). 'Disagreements, spin-offs and the evolution of Detroit as the capital of the US automobile industry'. *Management Science,* vol. 53, no. 4., pp. 616-631.

Kloosterman, R.C. (2001). *Ruimte voor reflectie* (oratie). Amsterdam: Universiteit van Amsterdam.

Koster, S. (2006). *Whose child? How existing firms foster new firm formation. Individual start-ups, spin-outs and spin-offs.* Groningen: Rijksuniversiteit Groningen, Faculteit Ruimtelijke Wetenschappen.

Kotler, P. (1995). *Marketing for non profit organizations.* Englewood Cliffs: Prentice-Hall.

Kotler, P., Haider, D. & Rein, I. (2002). *Marketing Places.* New York: The Free Press.

Krumme, G. (1969). 'Towards a geography of enterprise'. *Economic Geography,* vol. 45, no. 1, pp. 30-40.

Landes, D.S. (1998). *The Wealth and Poverty of Nations,* [edition 2007]. London: Abacus, London.

Leamer, E. & Storper, M. (2001). 'The Economic Geography of the Internet Age'. *Journal of International Business Studies,* vol. 32, no. 4, pp. 641-665.

Leone, R.A. & Struyk, R. (1976). 'The Incubator Hypothesis. Evidence from Five SMSAs'. *Urban Studies,* vol. 13, no. 3, pp. 325-331.

Levin, I. et al. (2002), 'A Tale of Two Pizzas. Building up from a basic Product versus scaling down from a fully-loaded Product'. *Marketing Letters,* vol. 13, no. 4, pp. 335-344.

Linders, G.J.M., Burger, M.J. & Oort, F.G. van (2005). 'A rather empty world. The many faces of distance and the persistent resistance to international trade'. *The Cambridge Journal of Regions, Economy and Society,* vol. 1, pp. 439-458.

Lösch, A. (1954). *The Economics of Location* (second revised edition, vertaald uit het Duits). New Haven: Yale University Press.

Manshanden, W.J.J. (1996). *Zakelijke diensten en regionaal economische ontwikkeling. De economie van nabijheid.* Amsterdam: Universiteit van Amsterdam.

Markoff, J. (2009), 'Searching for silicon valley'. The New York Times, april 16, 2009. [consulted] 21 maart 2012 [through] http://travel.nytimes.com/2009/04/17/travel/escapes/17Amer.html?pagewanted=all.

Marshall, A. (1920), *Principles of Economics. An Introductory Volume.* Londen: Macmillan.

Martin, R. (1999). 'The New "Geographical Turn" in Economics'. *Cambridge Journal of Economics,* vol. 23, pp. 63-91.

McCann, P. & Ortega-Argiles, R. (2011). *Smart specialization, regional growth and applications to EU cohesion policy* [Economic Geography Working Paper]. Groningen: Rijksuniversiteit Groningen, Faculteit Ruimtelijke Wetenschappen.

McCann, P. (2011). 'International business and economic geography. Knowledge, time and transactions costs'. *Journal of Economic Geography,* vol. 11, no. 2, pp. 309-317.

McNee, R.B. (1960). 'Toward a more humanistic economic geography. The geography of enterprise'. *Tijdschrift voor Economische en Sociale Geografie,* vol. 51, pp. 201-205.

Meester, W.J. (1999). *Subjectieve waardering van vestigingsplaatsen door ondernemers* (dissertatie). Groningen: Rijksuniversiteit Groningen.

Meester, W.J. & Pellenbarg, P.H. (2006). 'The spatial preference map of Dutch entrepreneurs. Subjective rating of locations, 1983, 1993 and 2003'. *Tijdschrift voor Economische en Sociale Geografie,* vol. 97, no. 4, pp. 364-376.

Metz, T. (2005), 'Creativiteit is geld waard. Terwijl de economie cultureel wordt, wordt de cultuur commercieel'. NRC Handelsblad, 18 februari 2005. [consulted] 12 maart 2012 [through] http://www.academic.lexisnexis.nl/.

Myrdal, G. (1957). *Economic theory and underdeveloped regions.* Londen: Druckwort.

Nelson, R.L. (1958). *The selection of retail locations.* New York: Dodge.

Nelson, R.R. & Winter, S.G. (1982). *An evolutionary Theory of Economic Change*. Cambridge: Harvard University Press.

Nieuwsblad Transport (2010). 'Beginnen met niets, dat trekt'. *Nieuwsblad Transport*, no. 17, p. 15, 29 april 2010.

Nooteboom, B. (1994). *Management van partnership in toeleveren en uitbesteden*. Schoonhoven: Academic Service.

North, D.C. (1989). 'Institutions and economic growth. An historical introduction'. *World Development*, vol. 17, no. 9, pp. 1319-1332.

North, D.C. (1955). 'Location Theory and Regional Economic Growth'. *Journal of Political Economy*, vol. 63, no. 3, pp. 243-258.

O'Brien, R. (1992). *Financial integration. The end of geography*. London: Royal Institute of International Affairs.

Oldenboom, E. (2005). *Costs and Benefits of major sports Events. A Case-Study of Euro 2000*. Amsterdam: Meerwaarde.

Oort, F. van (2001). *Agglomeration, economic growth and innovation*. Rotterdam: Erasmus Universiteit.

Oort, F. van, et al. (2007). *Verhuizingen van bedrijven en groei van werkgelegenheid*. Den Haag: Ruimtelijk Planbureau.

Organization for Economic Corporation and Development (2011), *Housing and the economy. Policies for renovation*. Parijs: OECD.

Parker, P. (2000). *Physio Economics. The basis for long term economic growth*. Cambridge: MIT Press.

Pellenbarg, P.H. (1985). *Bedrijfslokatie en ruimtelijke cognitie*. Groningen: Rijksuniversiteit Groningen, Faculteit der Ruimtelijke Wetenschappen.

Pellenbarg, P.H. & Meester, W.J. (2005). *Regional marketing to change regional images. The example of the Groningen campaign* (paper presented to the Academy of Marketing conference). Dublin, 5-8 juli 2005.

Pellenbarg, P.H., Meester, W.J. & Dijkema, J. (2005). 'Het image van de Friese steden. Subjectieve waardering van ondernemers voor vestigingsplaatsen in Friesland'. *It Beaken, Tydskrift fan de Fryske Akademy*, vol. 67, no. 2, pp. 79-100.

Pellenbarg, P.H., Steen, P. van & Wissen, L. van (red.) (2005). *Ruimtelijk aspecten van de bedrijvendynamiek in Nederland*. Assen: Van Gorcum.

Pen, C. (2002) *Wat beweegt bedrijven? Besluitvormingsprocessen bij verplaatste bedrijvenI*. Groningen: Rijksuniversiteit Groningen.

Perroux, F. (1950). 'Economic space. Theory and applications'. *The Quarterly Journal of Economics*, vol. 64, no. 1, pp. 89-104.

Porter, M. (1990). *The competitive advantage of nations*. New York: The Free Press.

Pred, A.R. (1967). 'Behavior and location. Foundations for a geographic and dynamic location theory'. *Studies in Geography*, vol. 27. Lund: Lunds Universitet.

Reformatorisch Dagblad (1990) 'Nieuw hoofdkantoor van PTT in Groningen betekent einde aan spreiding van rijksdiensten'. *Reformatorisch Dagblad*, 12 juni 1990. [consulted] 22 maart 2012 [through] http://www.digibron.nl/search/share.jsp?uid=00000000012dfcecbddef222a35e8e40&sourceid=1011].

Ricardo, D. (1817). *On the Principles of Political Economy and Taxation.* [consulted] 10 april 2012 [through] www.econlib.org/library/Ricardo/ricP.html

Ritchie, J.R.B. (1984). 'Assessing the Impact of Hallmark Events. Conceptual and Research Issues'. *Journal of Travel Research,* vol. 23, no. 1, pp. 2-11.

Rodrigue, J-P. (2011). 'The geography of transport systems'. [consulted] 15 november 2011 [through] http://people.hofstra.edu/geotrans/eng/ch2en/conc2en/locationfactors.html

Rugman, A.M. (2005). 'Rethinking International Management Theory in a World of Regional Multinationals'. *Internalization, International Diversification and the Multinational Enterprise: Essays in Honor of Alan M. Rugman* [A. Verbeke, ed.]. Research in Global Strategic Management, vol. 11, pp.165-202. Bingley: Emerald Group Publishing Limited.

Scott, A.J. (1988). 'Flexible production systems and regional development. The rise of new industrial spaces in North America and western Europe'. *International Journal of Urban & Regional Research,* vol. 12, no. 2, pp. 171-186.

Simon, H.A. (1960). *The new science of management decision.* New York: Harper & Row.

Smidt, M. de (1975). *Bedrijfsstructuur en arbeidsmarkt in een ruimtelijk kader, met een toepassing op de Drechstreek en het Noord-Merwedegebied.* Utrecht: Rijksuniversiteit Utrecht.

Smith, A. (1776). *The Wealth of Nations.* [consulted] 12 december 2011 [through] http://www.online-literature.com/adam_smith/wealth_nations/.

Smith, A. (1904). *An Inquiry into the Nature and Causes of the Wealth of Nations by Adam Smith,* ***vol. 1*** (edited with an Introduction, Notes, Marginal Summary and an Enlarged Index by Edwin Cannan). London: Methuen.

Snijders, H., Jacobs, D. & Vrolijk, H. (2007). *De economische kracht van agrofood in Nederland.* Den Haag: Stichting Maatschappij en Onderneming.

Sombart, W. (1907). Der Begriff der Stadt und das Wesen der Städtebildung. *Archiv für Sozialwissenschaft und Sozialpolitik,* vol.25, pp. 1-9.

Stec (2001, and more recent years). *Logica in locatiepatronen.* Stec Groep: Arnhem.

Storper, M. & Walker, R. (1989). *The Capitalist Imperative. Territory, technology and industrial growth.* New York: Basil Blackwell.

Storper, M. (1997). *The regional world. Territorial development in a global economy.* New York: The Guilford Press.

Taylor, P, (2001), 'Urban hinterworlds. Geographies of corporate service provision under conditions of contemporary globalization'.*Geography,* vol. 86, no. 1, 2001, pp. 51-60.

The World Bank (2009). *World Development Report 2009. Reshaping Economic Geography.* Washington D.C.: The World Bank

Tordoir, P. (2001), *Marktwerking en ruimtelijke organisatie.* Amsterdam: Vossiuspers, Universiteit van Amsterdam.

Transparency International (2011). *Corruption perceptions index 2011.* [consulted] 5 april, 2012 [through] http://cpi.transparency.org/cpi2011/.

United Nations Industrial Development Organization (2012). *The UNIDO Cluster/Network Development Programme. The Italian Experience of Industrial Districts.* [consulted] 21 maart 2012 [through] http://www.unido.org/index.php?id=o4310.

Vernon, R. (1966). 'International Investment and International Trade in the Product Cycle'. *The Quarterly Journal of Economics,* vol. 80, no. 2, pp. 190-207 & 225-267.

Walker Smith, J. (2011). 'The imperative of brand ritual'. *Marketing Management,* vol. 20, no. 2, pp. 48.

Wall, R.S., Burger, M.J. & Knaap, G.A. van der (2011). 'The geography of global corporate networks. The poor, the rich, and the happy few countries'. *Environment and Planning A,* vol. 43, no. 4, pp. 904-927.

Wall, R.S. & Knaap, G.A. van der (2011). 'Sectoral Differentiation and Network Structure Within Contemporary Worldwide Corporate Networks'. *Economic Geography,* Vol. 87, no. 3, pp. 267-308.

Wiemer, G. (2011). *Evolutie van het toeristisch product van de gemeente Ede.* Utrecht: Universiteit Utrecht, Faculteit Geowetenschappen.

Williams, D. (1995). 'Driven by Small Firms, Italy's Economy Thrives; Northern Manufacturers Beat the Odds. *The Washington Post,* 17 oktober 1995, p. A31.

Williamson, O.E. (1975). *Markets and hierarchies. Analysis and antitrust implications.* New York: The Free Press.

Wintjens, R. (2001). *Regionaal economische effecten van buitenlandse bedrijven. Een onderzoek naar verankering van Amerikaanse en Japanse bedrijven in Nederlandse regio's.* NGS 286. Utrecht: Universiteit Utrecht, Faculteit Ruimtelijke Wetenschappen.

Wolpert, J. (1963). *Decision-making in Middle Sweden's farming. A spatial, behavioral analysis,* Madison: University of Wisconsin.

Register